ORGANIC REACTIONS IN ELECTRICAL DISCHARGES

KHIMICHESKIE REAKTSII ORGANICHESKIKH PRODUKTOV V ELEKTRICHESKIKH RAZRYADAKH

ХИМИЧЕСКИЕ РЕАКЦИИ ОРГАНИЧЕСКИХ ПРОДУКТОВ В ЭЛЕКТРИЧЕСКИХ РАЗРЯДАХ

ORGANIC REACTIONS IN ELECTRICAL DISCHARGES

Edited by N. S. Pechuro

Department of Petrochemical Synthesis
Lomonosov Institute of Fine Chemical Technology
Moscow

Translated from Russian

 Springer Science+Business Media, LLC 1968

The original Russian text, published by Nauka Press in Moscow in 1966
for the Experimental Research Institute for Metal-Cutting Machinery of
the Ministry of Machine Tool and Instrument Production and for the Aca-
demy of Sciences of the USSR, has been corrected by the editor for this
edition.

Химические реакции органических продуктов
в электрических разрядах

Library of Congress Catalog Card Number 68-28093

ISBN 978-1-4899-2713-2   ISBN 978-1-4899-2711-8 (eBook)
DOI 10.1007/978-1-4899-2711-8

© 1968 Springer Science+Business Media New York

Originally published by Consultants Bureau in 1968

PREFACE

Recently interest has arisen in the use of electrical discharges to effect chemical reactions. Various types of discharges enable us to alter the conditions of a process over a wide range, and often to combine the high electron "temperatures" necessary to activate certain reactions with the low molecular temperatures of the medium used. This makes it possible to simplify the techniques and apparatus required for a process, and to "quench" the products, thus preventing them from decomposing again.

Some work on the use of electrical discharges has already found practical applications. These include the preparation of oxides of nitrogen, hydrocyanic acid, ozone, and hydrogen peroxide, thickening of oils, synthesis of acetylene and its homologs, hydrogenation and dehydrogenation of various oils and animal fats, etc.

In the heavy organic-synthesis industry, lower olefins and acetylenes are used on a large scale as raw materials for producing many valuable products. Methods of producing ethylene, propylene, and butylenes from gaseous and liquid hydrocarbon feedstocks have been widely adopted. However, the production of acetylene — which is an important and sometimes indispensable initial product — is mainly based on calcium carbide. Despite a number of recent improvements, this method of acetylene production suffers from high electric power consumption and expensive raw materials, requires several stages, and produces large amounts of waste products. Various lines are being followed in the attempt to find new, more sophisticated means of producing acetylene. After much research, the most promising methods to emerge have been thermooxidative and thermal pyrolysis, decomposition in streams of various heat-transfer media, and also decomposition of liquid, vaporized, and gaseous hydrocarbon feedstocks in electrical discharges and plasmas. For producing acetylene from hydrocarbon feedstocks, these last two methods (in discharges and plasmas) are exceedingly promising; they permit the processes to be so organized that large amounts of energy can be expended in relatively small reaction vessels. For this reason, research in this field is not only of theoretical interest, but also has much practical value.

The articles in this symposium deal with certain theoretical aspects of the mechanism of formation of acetylene in an electrical discharge, with the development of various alternative techniques for the processes, and also with some aspects of the technical and economic efficiency and prospects for development of methods of acetylene production.

One may hope that this book, which is the first attempt to correlate work on the use of electrical discharges to produce acetylene, will be of use to engineers and scientists who specialize in the chemical technology of organic products.

N. S. Pechuro

CONTENTS

ACETYLENE

N. P. Fedorenko and E. P. Shchukin

The economic importance of acetylene lies in its two main uses — for flame welding and metal cutting — and for the synthesis of many chemical products. In 1964, acetylene from calcium carbide was used in about equal amounts for metalworking and for organic synthesis.

The technical importance of welding is rapidly increasing. Welding has permitted marked improvements in mechanical and structural engineering, and has reduced the consumption of labor and metal. It has given rise to radically new economic solutions to constructional problems. Successive cutting and welding has enabled factories to fabricate large machine components which could not otherwise be moved to the site. Welded seams fasten millions of dwellings and factory buildings, thousands of kilometers of pipelines and railroad tracks, huge tanks for storing and transporting liquids and gases, and gigantic motor shafts.

There is an ever-growing need for gaseous heat sources for metal welding and cutting; this need will continue to grow with the development of heavy industry, transport, and building.

Acetylene occupies an important place as a source of high-temperature flame jets for metal cutting and welding. Its chief competitor is the electric arc. Electrical welding and cutting of metals is used instead of acetylene primarily in stationary plants, i.e., in factory shops in large towns and on large building sites. However, acetylene remains irreplaceable in newly developing regions, new building projects, and repair work.

In the flame treatment of metals, calcium carbide is in competition with acetylene from cylinders. It is clearly more economical to fill the cylinders with the cheapest possible acetylene. In present-day planning, carbide raw materials are yielding place to hydrocarbon raw materials in technical and economic efficiency. Thus, as plants for producing acetylene from hydrocarbon stocks are brought into use, they will be increasingly used for filling cylinders.

Carbide acetylene remains valuable for flame treatment of metals primarily because of the relative ease of transportation of carbide in comparison with cylinders, and because of delays in adopting methods for producing acetylene from hydrocarbon stocks.

According to preliminary estimates, the consumption of acetylene in metalworking in the USSR in 1970 will be 1.5 times the consumption in 1965, while that in chemical synthesis will be almost twice as great in 1970 as in 1965.

From the end of the nineteenth century to the nineteen-forties, acetylene was the most important intermediate product in the organic-synthesis industry. But then this industry began a new phase of prosperity as modern methods of petroleum processing were developed, and huge supplies of ethylene, propylene, butylene, and other similar products of the liquid-fuel and lubricant industry were placed at the disposal of chemists.

In a comparatively short time — 15-20 years — acetylene yielded pride of place to ethylene in such important syntheses as the production of plastics, solvents, and resins for synthetic fibers.

Countries with more limited means for petroleum refining used more acetylene from calcium carbide. The absolute value of carbide production did not decrease, because it depended on the total organic-synthesis production level. This being so, the following were the figures for calcium carbide production in 1962:

USA .... 980,000 tons
Federal German Republic ~1,000,000 tons
France .... 430,000 tons
UK. .... 250,000 tons
Japan .... 1,100,000 tons

There was a rapid increase in the proportion of acetylene produced from hydrocarbon crude in the capitalist countries. In 1957, less than 15% of the total was produced from hydrocarbons. In 1960-1961, this fraction increased:

	%
USA.	to 25
Federal German Republic	33
Italy	60

It is expected that by 1970 the proportion of acetylene produced from hydrocarbon crude will increase to 55-60%, and by 1975 to 70%.

In prewar Germany, and in the first postwar years in the Federal German Republic, organic synthesis was based mainly on coal and carbide acetylene. Only in 1955 did the West German chemical companies begin to speed up the creation of a petrochemical industry. The synthetic-fuel plants, which had previously supplied raw material for chemical processes, could not compete with petroleum, and cut down their activity, while the gas-coke industry was already unable to supply the required amounts of raw material.

Calcium carbide is now also beginning to lose its position as a source of acetylene for organic synthesis. Until 1960, the Federal German Republic was the largest producer of this product, but was beginning to use more and more petrochemical acetylene and ethylene. At present the latter is widely used. While the ratio of acetylene to ethylene used in the chemical industry was 1 : 0.4, in 1962 it already was 1 : 1 (Table 1).

In the total world consumption of acetylene for organic synthesis, the share of carbide fell from 72% in 1957 to 65% in 1961. It is considered that acetylene from petroleum and natural gas will take an even greater share of the requirements, because increasingly economic methods are being devised for its production.

Olefins for organic synthesis are produced in Italy mainly by cracking light gasolines. In 1962, the total quantity of ethylene used was 350,000 tons, and of propylene 300,000 tons per annum. As olefin production increased, the value of carbide as a source of synthesized acetylene correspondingly fell. Its production fell from 320,000 tons in 1958 to 290,000 tons in 1960.

In Britain, as compared with other West European countries, the earlier development of petrochemistry meant that acetylene was used as only a small proportion of the raw material for organic synthesis; in 1959, 24% of this was acetylene and 76% was ethylene (Table 2).

In Japan, together with petroleum and natural gas, large quantities of calcium carbide were also used as the raw material for synthesis in order to satisfy the needs of the internal market: in 1955, of 674,000 tons of calcium carbide used, 363,000 tons (54%) were converted to calcium cyanamide and 182,000 tons were used (27%) in the organic synthesis industry; in 1963, 310,000 tons (22%) were used in cyanamide production, and 800,000 tons (58% of the total calcium carbide) in synthesis, and only about 20% for welding and metal cutting.

The output of the Japanese carbide works became the highest in the world — about 2.5 million tons per year. But their productivity was at least 2.5 times less, owing to intermittence in the electricity supply from hydro-stations which during much of the year operate below full capacity.

TABLE 1. Federal German Republic. Consumption of Acetylene and Ethylene in Organic Synthesis Production (in thousands of tons)

	1957	1959	1961
Acetylene	185	247	280
Ethylene.	75	165	270
Ratio of acetylene to ethylene.	1:0.4	1:0.67	1:0.96

TABLE 2. Great Britain. Consumption of Raw Materials for Production of Organic Synthesis Products

	1949		1953		1962	
	tons $\cdot 10^3$	%	tons $\cdot 10^3$	%	tons $\cdot 10^3$	%
Coal tar . , ...	175	34.7	230	30.9	540	24.4
Acetylene (from carbide)	70	13.8	80	10.7	150	6.8
Synthesis gas (coke). .	55	10.9	70	9.4	90	4.1
Fermentation products	160	31.7	170	22.8	30	1.3
Petroleum	45	8.9	195	26.2	1400	63.4
Total	505	100.0	745	100.0	2210	100.0

TABLE 3. USA. Output of Organic Products and Distribution with Regard to Sources of Raw Material in 1962 (in thousands of tons)

	Vinyl chloride	Vinyl acetate	Acrylo-nitrile	Trichloro-ethylene
Total output.	742	201	203	220
Based on acetylene				
from carbide	225	125	36	190
from petroleum.	175	113	91	—
Based on ethylene and				
other hydrocarbons.	342	23	76	30
Ratio of output of carbide-based acetylene to total output, %	30	48	18	87

In the organic synthesis industry of the USA, carbide acetylene is used as well as petroleum. However, the competition between acetylene and ethylene as raw material for organic synthesis ended with victory for the cheaper and more available ethylene: its production rapidly rose from 2,500,000 tons in 1959 to 3,600,000 tons in 1962. It is expected to rise by 500,000 tons in the next few years.

In 1963, over 20 carbide works were in operation in the USA, their total output being about 700,000 tons.

The ratio of acetylene to ethylene as synthesis material was 1:4 in 1957 and 1:8 in 1962. The total consumption of acetylene and ethylene for synthesis was 2,180,000 tons in 1959: acetylene constituted 13.6%, and ethylene 86.4%. In Western Europe, where the petrochemical industry is just starting, the proportions of acetylene and ethylene were 61.7 and 38.3% in the Federal German Republic, and 61.6 and 38.4% in France.

In 1962, the USA produced 79% of its acetylene from carbide; this gas was used mainly to make neoprene, vinyl chloride, vinyl acetate, acrylonitrile, and trichloroethylene (Table 3).

Until 1957, the outputs of ethylene and acetylene in the USA, as expected, were 3 and 1.3 million tons. Such a marked change in the structure of the acetylene industry was due to economic causes. The production of ethylene from petroleum refining products and gaseous byproducts costs only two-thirds as much as its production from calcium carbide.

The thermal, electrothermal, and thermooxidative methods of converting methane and petroleum hydrocarbons to acetylene are cheaper than carbide production. But, as a rule, ethylene is more cheaply produced from hydrocarbon feedstock. The recently introduced methods of joint production of acetylene and ethylene make it possible to adopt the same rate for both products, but even then ethylene retains its advantage, because its further processing to polymer materials is usually cheaper, and possibilities for this processing are more varied and readily available.

However, there are certain syntheses in which ethylene cannot yet replace acetylene, or can replace it only partly — trichlorethylene, "Nairit" rubber [neoprene], vinylacetylene, and vinyl chloride. Methods of producing acetylene continue to improve, and it is becoming cheaper. It is produced from such cheap and easily available raw materials as methane from natural gas, and heavy petroleum fractions which are unsuitable for fuel production. Not unimportant is the reliability of the well-tried carbide process, which yields a high-quality product with a minimum content of impurities which might exert an unfavorable influence on organic synthesis processes.

The present ascendance of ethylene and its derivatives will certainly not eliminate the need for the development and expansion of acetylene production.

In the last decade, the most important advance in the economics of acetylene production has been the adoption of hydrocarbons as raw material.

Many methods have been suggested for obtaining acetylene from hydrocarbons. They all differ either in the type of raw material used, or in the method of creating a high temperature in the reaction zone. In classifying these methods it is convenient to consider first the cracking of gaseous and liquid hydrocarbons. Decomposition can be effected in an electrical discharge with regeneration of heat from the gas, or by burning part of the raw material in the reaction zone or on the reactor walls with a forced feed of commercial or atmospheric oxygen.

The organic synthesis industry of the USSR uses practically all the above new methods of acetylene production. But at the moment there are three main types of process being successfully adopted for the treatment of hydrocarbon stock: oxidative pyrolysis of natural gas, electrocracking of natural gas, and pyrolysis of liquid hydrocarbons.

Since the treatment of ethylene has recently become cheaper, production plans for acetylene have been revised and reduced. However, the adoption of the new processes has been delayed, and carbide production still represents more than half of the total capacity.

In Soviet conditions, electrocracking of natural gas has not come up to expectations. In addition to purely technical difficulties, its expansion is impeded by economic conditions. The chief of these is the relatively high cost of electric power in comparison with that of natural gas. Acetylene production by electrocracking can only be profitable when it is based on extremely cheap power from large hydroelectric stations, i.e., in only a limited number of regions of the USSR. Production of electrical power from thermal stations fired by gas does not alter the cost ratio sufficiently to restore the advantage to electrocracking. The gas consumed in power production and the energy losses during cracking remain greater than those in thermal and thermooxidative cracking.

If 0.4 kg of standard fuel is expended on each kilowatt-hour of electrical energy (or at the rate of 0.56 kWh/m^3 oxygen), the process of thermooxidative pyrolysis consumes about 10.5 tons of standard fuel per ton of acetylene formed, allowing for power and technical (methane for pyrolysis) requirements.

TABLE 4. Consumption Indices of Various Methods of Acetylene Production

Index	Carbide acetylene	Acetylene from thermooxidative pyrolysis of natural gas
Raw material	36	112
Power	65.5	37
Total	101.5	149
By-products*	1.5	65
Total/ton acetylene	100	84

* Subtracted.

At the State Institute for the Nitrogen Industry, calculations have been performed on the expenditure of raw material and power (in terms of standard fuel, in relative units); the results are listed in Table 4 (as percentages).

In thermooxidative pyrolysis an important part is played by the by-product synthesis gas — a mixture of carbon monoxide and water in a ratio close to that required for the production of ammonia and methanol. If this gas is burned, its calorific value enables it to supply about 45% of the thermal potential of the raw material and power used. It is very much more efficient as a source of chemicals than as a fuel.

Meanwhile, electrocracking of methane and pyrolysis of low-octane gasoline fractions to ethylene and acetylene have yielded place to thermooxidative pyrolysis of natural gas. In the first case, the heavy consumption (10,000 to 12,000 kWh/ton of acetylene) of electric power, the difficulty of removing soot and by-products, and the unreliable operation of the cracking plants make the process not so much unprofitable as un-stable. This instability ultimately leads to additional expenses, while contamination with organic by-products makes it difficult to get good products for further syntheses. These technical difficulties can apparently be overcome, and do not rule out further development of chemical production based on the electrical cracking of methane. The high power consumption of the process will always limit its use in regions of the USSR where the fuel—energy balance is under stress or where expensive fuel must be burned in power-station boilers. On the other hand, the presence of cheap electric power offers attractive prospects which cause us to pay continued attention to the development of the methane electrocracking method. This method may play an increased part in the total production of acetylene in the future, thanks to its use in those regions which have natural-gas deposits or large hydroelectric sources.

Oxidative pyrolysis of light liquid hydrocarbons to ethylene and acetylene has the advantages of stable feedstock composition and savings by joint production with other products. However, neither in the Soviet Union nor abroad have the expected yields been achieved. Here, as in the case of electrocracking, possibilities remain for technical and economic improvement of the process. However, no rapid and cheap solution has yet been found to the problem of obtaining cheap acetylene by pyrolysis of low-octane gasoline fractions.

In all the processes except the carbide one, the first stage of obtaining acetylene by decomposing hydrocarbons requires equipment to heat the feedstock to a high temperature, followed by rapid cooling of the reaction gases.

During partial oxidation, part of the methane burns and heats the rest, which is cracked, and the temperature of the reaction zone reaches about 1500°C. At this temperature the maximum yield of useful product is about 75%. Reduction of the partial pressure of feedstock by feeding in hydrogen can give more than 80% conversion — a figure which many investigators consider to be the limit. The combustion products (CO and CO_2) cause additional difficulties in purifying the gas and increase the capital costs of the plant.

An important part of the incomplete-oxidation process is coke formation. As the efficiency of the cracking process decreases, a greater amount of methane is converted to coke, which can choke the reactor or the purification and cooling plant. However, by strict observation of the process conditions coke formation can be suppressed so far that the apparatus will work stably for a long time.

To achieve high conversion of hydrocarbons to acetylene and to improve the thermal balance of the process, high-pressure pyrolysis is being studied. It is expected that this will reduce the heat consumption by about 10%.

In electric-arc processes for acetylene production, the gas is heated to several thousand degrees. In these conditions there is intense ionization, and acetylene is one of the reaction products. Although the arc heats up the raw gas (methane) rapidly, a small part of its energy is used for the reaction. This causes an increased consumption of electric power.

A study of the process kinetics and control of the temperatures in the feedstock-preparation, reaction, and quenching zones permits the power consumption to be reduced by a factor of 1.5-2.

Much interest attaches to experiments on the decomposition of liquid hydrocarbon feedstock in arc discharges. In contrast to the other methods, here we are using a fundamentally different process structure — instead of feeding raw material into the high-temperature reaction zone created by heating gas in an electric arc or plasma-burner flame, we strike the arc directly in the liquid hydrocarbon stock. The advantage of this method is mainly that the thermal fluxes are "closed." The heat from the arc can reach only the surrounding layer of stock, heating it and preparing it for the reaction. The gas mixture formed is rapidly cooled by the surrounding liquid, and so is naturally quenched. By controlling the depth of the liquid feedstock and, if necessary, by using a simple device for breaking up gas bubbles rising to the surface, we can easily attain the required degree of cooling of the products of the first stage of cracking, which are the richest in acetylene. The quenching time is automatically reduced to a minimum: this, of course, is very important in order to prevent secondary processes which might convert part of the acetylene into by-products.

When the process is organized in this way, there is no need for special structures, at least for the first stage of cooling of the products; heating of the stock before it reaches the reaction zone becomes automatic and removal of soot is facilitated. The soot is washed out of the gas and remains in the liquid feedstock layer. Thus, the trapping of soot from the gaseous products is replaced by separation of soot from the liquid feedstock, which is simpler and cheaper. We must not, however, lose sight of the fact that for some purposes which are especially sensitive to the presence of soot in the acetylene, the operation of fine gas purification must be retained.

In producing acetylene from hydrocarbon stock, much of the cost is due to the necessity of concentration, because the acetylene content of the gas is from 6 to 15% by volume. When liquid hydrocarbons are decomposed in electrical discharges, we can obtain a gas which contains about 30% by volume of acetylene. This reduces the concentration costs.

Not less important is the universal nature of this process: the raw materials can be various types of liquid hydrocarbon, from light gas condensates to boiler fuels (petroleum residues). The acetylene yield goes down as the stock composition gets heavier, while the power consumption rises; however, this is accompanied by a transition to cheaper and more readily available petroleum products, which largely compensates for the reduction in process efficiency.

In theory, by choice of the most efficient raw material (in which aromatics predominate) and minimization of energy losses, we can reduce the electric power consumption to 6-7 kWh per cubic meter of acetylene.

Since the reactor is relatively simple, and some product-cooling and gas-purification and concentration operations are eliminated, the capital cost of the plant is lower than that of other methods of acetylene production.

According to preliminary calculations, the cost of acetylene obtained by decomposing boiler fuel or light gas-condensate fractions in a high-voltage arc should be 20-30% lower than that of acetylene from high-temperature pyrolysis of low-octane gasoline.

Recently, in the Soviet Union [1] and elsewhere, research has been done on the use of plasma devices in the decomposition of various types of hydrocarbon stock for the purpose of producing acetylene. The intermediate medium is argon or hydrogen; the cracked product is mainly methane. Not excluded is the possibility of direct use of methane, without an intermediate medium, in specially constructed plasma burners. The latter method should simplify the technology of this process by eliminating the devices for separating and circulating argon or hydrogen.

It has been shown experimentally that in the case of decomposition of methane in argon, the consumption of electrical energy is about 15 kWh/m^3 acetylene. The acetylene concentration in the gas is about 8-10% by volume; the degree of conversion is 80%.

This process is still at the laboratory stage, but is of undoubted practical interest. Preliminary calculations show that the cost of acetylene produced by a plasma device may be about the same as that of acetylene produced by thermooxidative pyrolysis of methane.

This survey of current research and methods of acetylene production has shown that development of these processes, somewhat retarded in recent years owing to a period of industrial adoption, is again gathering pace, and shows promise of important successes.

Acetylene will continue to be a rival to ethylene as the raw material for organic synthesis, and this competition will lead to an overall improvement in the economics of this important field.

LITERATURE CITED

1.   Symposium, Kinetics and Thermodynamics of Chemical Reactions in Low-Temperature Plasmas, Izd. Nauka, Moscow (1965).

PYROLYSIS OF GASOLINE VAPOR TO ACETYLENE AND OLEFINS
IN A STEAM PLASMA

D. T. Il'in and E. N. Eremin

By the pyrolysis of the vapors of individual hydrocarbons and gasoline in a hydrogen plasma, high concentrations of unsaturated compounds have been achieved in the pyrolysis gas (56% by corrected volume) [1, 2]. It seemed of interest to use an energy transfer agent which would condense after the reaction, so as to bring the actual concentration of unsaturated compounds in the product gas near to the corrected value obtained when working with a hydrogen plasma. For this purpose we have studied the pyrolysis of gasoline vapor in a steam plasma.

We have shown [1] that the results of experiments on the pyrolysis of gasoline vapor in a steam plasma can be plotted, for example, at constant specific energy, as graphs of the reaction-product concentration and energy consumption versus the specific gasoline consumption (δ, in liters of gasoline liquid/liter of liquid water). Figure 1 is an example of such a graph.

The graphs of accumulation of the principal pyrolysis products and the curve of energy consumption versus δ have the same shapes as those found for gasoline pyrolysis in a hydrogen plasma [2, 3]. But, for the

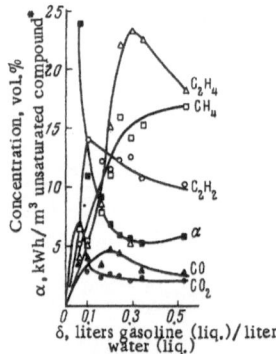

Fig. 1. Concentrations of products of gasoline pyrolysis in steam plasma and energy consumption α, versus specific gasoline consumption δ. U/V_{H_2O} = 0.67 kWh/liter water (liq.). (•) All volumes reduced to NTP.

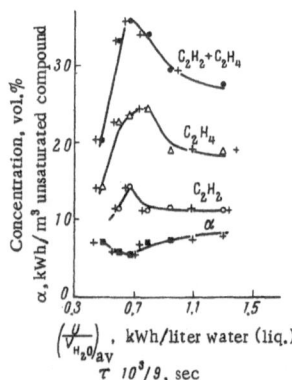

Fig. 2. Concentrations of products of gasoline pyrolysis in steam plasma and energy consumption ∞, versus specific energy of steam arc (U/V_{H_2O}) and time of contact in reaction zone τ.

8

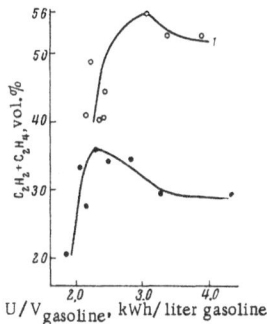

Fig. 3. Concentrations of unsaturated compounds in gases obtained by pyrolysis of gasoline in hydrogen and steam plasmas, plotted versus specific pyrolysis energy $U/V_{gasoline}$. Pyrolysis: 1) in hydrogen plasma; 2) in steam plasma.

same degree of gasoline conversion, the concentrations of unsaturated compounds are here much higher, because, on leaving the pyrolysis chamber, the steam is condensed and separates out from the pyrolysis products. For example, when the specific energy U/V_{H_2O} was 0.67 (Fig. 2), the concentration of unsaturated compounds in the final gas reached 36 vol.%, of which 10.7 vol.% was acetylene. From the measured quantity of gasoline carried away in the water, we found that the overall degree of conversion of gasoline to gaseous products was close to unity. With this pyrolysis method we get a number of new products: CO (up to 7 vol.%), CO_2 (up to 7 vol.%), and O_2 (up to 2 vol.%).

The relative heights of the maxima of unsaturated-compound concentrations and the relative positions of the optimum process parameters on the horizontal axis are roughly the same as in the experiments on gasoline pyrolysis in a hydrogen plasma; this shows that the two methods are similar. The highest acetylene concentrations correspond to specific gasoline consumptions around 0.2 liters of gasoline (liq.)/liter of water (liq.); the maximum ethylene and methane concentrations and minimum energy consumption α correspond to $\delta = 0.3$-0.4 liter/liter.

Figure 2 is based on six graphs of the above type with different values of U/V_{H_2O}: it plots the positions of the extrema on graphs like those of Fig. 1. The maximum process indices are here given in the form of functions of two variables — the specific arc energy (U/V_{H_2O}) and the reaction time (τ). As will be seen, each of the graphs plotted versus U/V_{H_2O} coincides with the corresponding graph plotted versus τ; the coefficient of proportionality between τ and U/V_{H_2O} is $10^3/9$ — the same as in the case of gasoline pyrolysis in a hydrogen plasma [2,3].

All the curves in Fig. 2 have clearly marked extrema in the region $U/V_{H_2O} = 0.7$ kWh/liter water (liq.). In this region the maximum C_2H_2 concentration is 14 vol.%, that of C_2H_4 is 24 vol.%, and that of $C_2H_2 + C_2H_4$ is 36 vol.%, while the minimum energy consumption is 5.3 kWh/m^3 $C_2H_2 + C_2H_4$. Here the minimum value of α is somewhat higher than that obtained with a hydrogen plasma (4 kWh/m^3 $C_2H_2 + C_2H_4$); this is due to the side reactions which produce CO and CO_2.

Figure 3 compares the data for the hydrogen and steam plasmas. Here we plot the overall maximum concentrations of unsaturated compounds in the pyrolysis gas (vol.%) versus the specific pyrolysis energy $U/V_{gasoline}$. For the case of pyrolysis in a hydrogen plasma, the concentrations of unsaturated compounds in the pyrolysis gases c_1 were computed from the formula

$$c_1 = c\,\frac{V_{\text{K}}}{V_{\text{H}_2}},$$

where c is the concentration of unsaturated compounds in the final gas, V_K is the volume of product gas per hour, liters/h, and V_{H_2} is the consumption of hydrogen in liters/h.

From Fig. 3 we see that the graphs are of similar shape: in particular, they each have a maximum — for the steam plasma this lies at $U/V_{gasoline} = 2.3$ kWh/liter of gasoline (liq.) and for the hydrogen plasma it lies at $U/V_{gasoline} = 3.1$.

The first of these maxima (36 vol.%) is lower than the second (56 vol.%) owing to the formation of CO and CO_2. However, with the same specific energy, corresponding to the first maximum, the results do not differ so widely — 36 as against 42 vol.%. If the specific pyrolysis energy for the steam plasma is further increased, the formation of CO and CO_2 is markedly increased, and the concentration of unsaturated compounds falls off.

As in the case of gasoline pyrolysis in a hydrogen plasma, in this process there is very little soot or tar in the reaction products. This, together with the high concentration of unsaturated compounds resulting from relatively low energy expenditure, suggests that the plasma-jet method may prove usable on a heavy industrial scale.

If the process described in this article were used on such a scale, we should probably get a gas containing about 50 vol.% of unsaturated compounds for an energy expenditure of at most 4 kWh/m^3.

Furthermore, our work has shown that the pyrolysis results are to some extent independent of the nature of the energy transfer agent: this makes it realistic to hope that a similarly high degree of conversion to unsaturated compounds may be obtained from hydrocarbons from the second stage of two-stage electric-arc pyrolysis with the heated cracking gases as heat-transfer agent.

SUMMARY

1. We investigated the pyrolysis of low-octane gasoline vapor to acetylene and olefins in a steam plasma formed by an electric arc at about 1.5 atm. The process was conducted so that hydrocarbons could not get into the discharge zone.

2. We found that, as in the case of pyrolysis in a hydrogen plasma, the gasoline — despite its complex composition — was mainly converted to acetylene, ethylene, and methane.

3. We got 100% overall conversion of gasoline to gaseous substances in the range of specific energies under study; there was very little soot or other solids in the reaction products.

4. On comparing the results obtained with hydrogen and steam plasmas, we found that the yield of unsaturated hydrocarbons was somewhat lower in the latter case; but owing to condensation of water after the reaction, the concentration of these compounds in the final gas was more than twice as high.

LITERATURE CITED

1.   Il'in, D. T., and E. N. Eremin, Zh. Obshch. Khim., 36 : 1560 (1962).
2.   Il'in, D. T., and E. N. Eremin, this collection, p. 11.
3.   Il'in, D. T., and E. N. Eremin, Vestn. Moskovsk. Univ., No. 2, 48 (1962).

PYROLYSIS OF SINGLE HYDROCARBONS OR GASOLINE VAPOR
BY MIXING WITH HYDROGEN PLASMA FROM AN ELECTRIC ARC

D. T. Il'in and E. N. Eremin

The maximum reaction temperature in purely thermal methods of producing acetylene and the olefins is limited by the combustion temperature of the fuel, 2500-3000°K.

In methods which involve an intermediate stage of heating up of a reaction surface (for example, Wulf's method [1]), the temperature is even lower. In processes based on direct contact between the hydrocarbon crude and the combustion products (Eastman's process [2]), the maximum pyrolysis temperature is 2500-2700°K.

Nevertheless, it is of practical and theoretical interest to consider pyrolysis at higher temperatures: in this region, the formation of acetylene becomes thermodynamically more favorable than decomposition of the original hydrocarbon to soot. The use of an electric arc offers practically unlimited promise in this direction.

We have carried out pyrolysis of gaseous hydrocarbons (n-hexane, n-heptane, n-octane, and iso-octane) and low-octane gasoline in a current of hydrogen plasma: the heat content of the hydrogen corresponded to an equivalent temperature of about 5000°C [3].

Simultaneously but independently of our work, the plasma jet method was used [4] abroad for the pyrolysis of methane in an argon plasma at a temperature (12,000°K) corresponding to 5000-6000°K in a hydrogen plasma with the same heat content.

Our work had an additional aim — to simulate the second stage of two-stage electric-arc pyrolysis. The idea of this method is to use the heat from the outgoing gases from methane cracking for converting heavy hydrocarbons (in particular, gasoline [5]) to acetylene and the olefins. Here the heavy hydrocarbons can partly replace the water which is usually added to the products of methane electrocracking so as to reduce the temperature rapidly (i.e., to "quench" the equilibrium). In our work, the gases from methane cracking — which, of course, are 50-60% hydrogen — were replaced by hydrogen.

In processing the results, from the range of hydrogen-arc specific energies under study we selected a value close to 1.3 kWh/m³ H₂.* In this case, the heat content of the hydrogen is practically the same as that of the gases from methane cracking at optimum specific energy (2.5 kWh/m³ CH₄).

From meter readings and analysis results, we found the following process parameters:

1. Current strength of hydrogen arc, I, amperes (dc).

2. Arc voltage, E, kV.

3. Arc power $U = EI$, kW.

---

*All volumes reduced to NTP — Translator.

4. Hydrogen flow rate, V_{H_2}, m^3/h.

5. Specific energy of arc U/V_{H_2}, kWh/m^3 H_2.

6. Flow rate of hydrocarbon (gasoline), V_{hyd} $(V_{gasoline})$, liters of hydrocarbon (gasoline) per hour.

7. Specific flow rate of hydrocarbon (gasoline), $\delta = V_{hyd}/V_{H_2}$ $(V_{gasoline}/V_{H_2})$, in liters of hydrocarbon (gasoline) per m^3 H_2.

8. Specific pyrolysis energy, U/V_{hyd}, kWh per liter of hydrocarbon (gasoline) (liq.).

9. Rate of formation of final gas, V_f, m^3/h.

10. Volumetric coefficient of expansion of gas as a result of the reaction, $\beta = V_f/V_{H_2}$.

11. Energy consumption rate, $\alpha = (U \cdot 100)/[V_f \cdot (\text{vol.% unsat. compound})]$.

12. Effective temperature of hydrogen plasma, T_{max}, °K; this is calculated [6] from the heat content of a volume V of hydrogen when energy U is imparted to it. (In this calculation we did not allow for the energy of dissociation of the hydrogen, because this is recovered owing to recombination of atomic hydrogen in the reaction zone.)

13. Final gas temperature, T_{min}, °K — in the case of pyrolysis of n-hexane; this was calculated from the energy balance of the hydrogen plasma by means of the heat capacity series for hydrogen and the reaction products, including the temperature in powers up to the fourth [6, 7].

14. The reaction time, τ, sec. This was found from the expression $\tau = V_r/V_f$, where V_r is the volume of the reaction channel.

15. The degrees of conversion of the hydrocarbons to acetylene ($\gamma_{C_2H_2}$), ethylene ($\gamma_{C_2H_4}$), and methane (γ_{CH_4}), in the case of pyrolysis of single hydrocarbons. These were calculated from the mass balance of the reaction.

It is very difficult to calculate the temperature and reaction time in the complicated conditions of plasma-jet synthesis [8]. In fact, in our case we have, not a single reaction temperature, but a whole range of temperatures, beginning with the maximum temperature close to the molecular—atomic temperature of the plasma, and ending with the minimum reaction temperature close to that of the final gas. Thus, the reaction time is also indeterminate, and this uncertainty is made worse by the fact that the reaction-zone volume is also essentially unknown: the same part of the reaction channel may be the reaction zone for some molecules and the quenching zone for others.

On this basis, and since (as already mentioned) the dissociation energy of the hydrogen molecule is recovered in the reaction, the maximum temperature T_{max} is a suitable parameter for comparing the results of different series of experiments and also results obtained with different types of plasma, e.g., argon [4] or steam [9].

Clearly more important for a consideration of the experimental results is the temperature T_{min} of the final gas: it is at this temperature, regardless of the differentiation of the reaction time, that practically all the reacting particles spend some time in the prequenching zone. Here, in calculating the reaction time, no account was taken of the thermal expansion of the gases.

The experimental data on gasoline pyrolysis in a hydrogen plasma can be plotted [10] as graphs of the pyrolysis product yield (vol.%) and energy consumption α (kWh/m^3 unsaturated compounds, acetylene + ethylene) versus the specific consumption of gasoline δ.

Chromatographic analysis of the pyrolysis products of gasoline and individual hydrocarbons revealed that, at medium and small values of δ, there is only a very small amount of higher compounds in the gas phase. At the highest values of δ used in our work, the total content of these compounds was under 1 vol.%. In discussing the results we therefore considered only those relations which refer to the formation of acetylene, ethylene, and methane. An example of these relations is shown in Fig. 1.

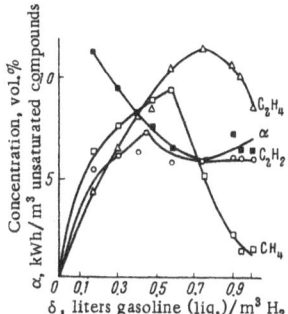

Fig. 1. Concentration of products of gasoline pyrolysis in hydrogen plasma, and energy consumption α, versus specific gasoline consumption δ. $U/V_{H_2} = 1.56$ kWh/m³ H₂.

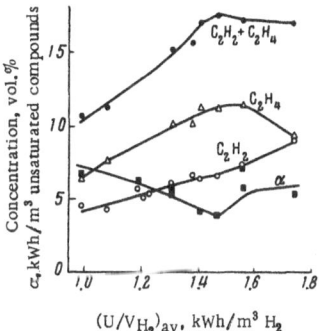

Fig. 2. Concentration of products of gasoline pyrolysis in hydrogen plasma and energy consumption α, versus specific energy of hydrogen arc (U/V_{H_2}).

All the analogous graphs have a similar shape: the maximum acetylene, ethylene, and methane concentrations are quickly reached, and then there is a slow fall, accompanied by an increase of specific gasoline consumption and an inverse relation for the energy consumption. A characteristic feature of the curves is that the maximum acetylene concentration is observed earlier than the maximum ethylene and methane concentrations, which approximately coincide with each other and with the minimum energy expenditure.

The optimum pyrolysis indices, according to graphs analogous to Fig. 1, correspond to the following values of δ: maximum acetylene concentration, 0.2-0.3; maximum ethylene concentration, 0.5-0.9; minimum energy expenditure, also 0.5-0.9 liters of gasoline (liq.) per m³ H₂.

On the whole, as we might have expected, there is a tendency for the maximum ethylene concentration and minimum energy expenditure to move toward higher δ as the specific energy increases.

It is interesting that, in the conditions described, the original hydrocarbon scarcely decomposes at all into its elements. This phenomenon, which is also observed during the pyrolysis of methane in an argon plasma [4], is of great practical importance, and distinguishes pyrolysis in plasma from pyrolysis in the arc itself, in which there is a great deal of decomposition to the elements. The overall conversion of gasoline to gaseous products, for the range of specific consumptions under study, is close to unity, and decreases appreciably only after passage through the maximum ethylene concentration.

Figure 2 is based on eight series of experiments similar to those plotted in Fig. 1, but with different specific energies. The curves plot the loci of the maximum concentrations of unsaturated compounds in the final gas, and of the minimum energy expenditures. The independent variable is the specific energy of the hydrogen arc, $U/V_{H_2}*$; in nearly every case, the average hydrogen consumption was close to the value 10 m³/h. We ignored those experiments in which the value of V_{H_2} deviated significantly from the mean. As the figure shows, when the specific energy is 1.45-1.55, the maximum ethylene concentration and the total of unsaturated compounds pass through maxima of 11.4 and 17.5 vol.%, respectively, and the minimum energy expenditure passes through a minimum of 4.0 kWh/m³ unsaturated compounds, equivalent to an expenditure of 54% of the arc's energy on acetylene and ethylene formation. At the same time, the maximum acetylene concentration continues to rise, and at the highest specific energies attained in the experiments it is 9 vol.%.

A further idea of the results of gasoline pyrolysis in a hydrogen plasma can be obtained from Fig. 3, which was plotted like Fig. 2, except that here the curve is the locus of the maximum corrected concentration

*In the experiments with $U/V_{H_2} = 1.56$ and 1.74 kWh/m³ H₂, the hydrogen was preheated to 900°C.

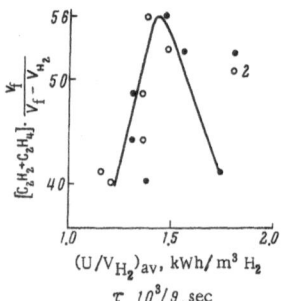

Fig. 3. Total corrected concentration of unsaturated compounds in products of gasoline pyrolysis in hydrogen plasma, plotted versus specific energy U/V_{H_2} (1) and contact time τ (2).

of the total unsaturated compounds, equal to the concentration of these compounds in the pyrolysis gases,* and that τ is used as an additional argument. As will be seen from the figure, this curve passes through a well-marked maximum at a specific energy of 1.47 kWh/m³, reaching 56 vol.% $C_2H_2 + C_2H_4$, which on average corresponds to 80-90% conversion of the original gasoline to acetylene and ethylene. It is interesting to note that this result is not achieved by other pyrolysis methods. Also noteworthy is the coincidence of the two graphs (that versus τ and that versus U/V_{H_2}): the ratio between the two arguments is $10^3/9$.

These results are not only interesting in their own right, but also confirm that it would be efficient to introduce a second pyrolysis stage in the electrocracking of methane.

Returning to Fig. 2, we see that, for a specific energy corresponding to simulation of this second stage (1.3 kWh/m³ H_2), we get about 15 vol.% of unsaturated compounds, of which about 6 vol.% is acetylene and the rest ethylene. We can expect that about the same amount of unsaturated products will be obtained from a second pyrolysis stage, working with methane and a specific energy of 2.5 kWh/m³ CH_4, in addition to 13 vol.% C_2H_2 and 1 vol.% C_2H_4 formed in the first stage. †

There are some publications [5, 11] on two-stage hydrocarbon pyrolysis in which the specific gasoline consumptions were higher than ours. In this respect these investigations are complementary.

Let us now consider the experimental results on the pyrolysis of individual hydrocarbons — n-hexane, n-heptane, n-octane, and iso-octane — in a hydrogen plasma. This work was performed with the same conditions as for gasoline, with specific arc energy 1.2-1.3 kWh/m³ H_2, and the results are plotted in Fig. 4 as a graph of the product concentrations and energy expenditure versus the specific expenditure of hydrocarbons δ, expressed in liters of hydrocarbon (liq.) per m³ H_2.

From this figure we see that, for all the pyrolysed hydrocarbons, the curves obtained have the same shape and the same relative positions of extrema as the analogous curves for the pyrolysis of gasoline. Despite the great difference between the raw materials, both gasoline (a mixture of 76 different hydrocarbons) and single compounds gave, in the conditions under study, the same products in approximately the same ratios. An exception to this rule is iso-octane (Fig. 4d); in this case, the maximum methane concentration (13.7 vol.%) is higher than that of ethylene (about 9 vol.%). However, the structure of the hydrocarbon has no effect whatever on the maximum acetylene concentration, which is 6.3 vol.% for both n-octane and iso-octane. Since the maximum acetylene concentration appears at a higher temperature ($\delta_{C_2H_2} < \delta_{C_2H_4}$), it is possible that the mechanism of acetylene formation is based on processes involving "deep" decomposition of the hydrocarbons, giving "more elementary" products than in the formation of ethylene. In any case, from the example of n-octane and iso-octane pyrolysis we can infer that the formation of acetylene probably involves a greater participation of the short methyl and methylene radicals. Ethylene is apparently formed mainly via larger molecular fragments by a single removal of radicals with two carbon atoms, while the unpaired radicals with one carbon atom go mainly to form acetylene and methane molecules; the higher the temperature, the more probable is the participation of these radicals in reactions which lead to the formation of acetylene. It is also possible that hydrogen atoms, supplied in excess by the hydrogen plasma (especially in the high-temperature region), play a not unimportant part in the formation of acetylene and ethylene.

---

* After subtraction of the original hydrogen.

† Our results for simulation of a second stage were confirmed by working with mixtures of CH_4 and H_2. With $U/V = 1.9$ kWh/m³ $CH_4 + H_2$ ($CH_4 : H_2 = 1:1$), in the second stage we obtained an additional 1.6 vol.% of unsaturated compounds. The initial material for the second stage was low-octane gasoline (A = 66).

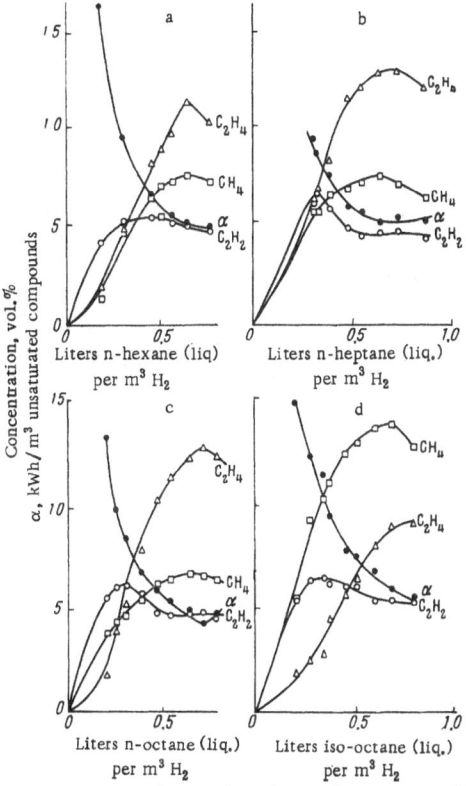

Fig. 4. Concentration of pyrolysis products and energy expenditure (α) versus specific hydrocarbon consumption (δ). Pyrolysis of a) n-hexane; b) n-heptane; c) n-octane; d) iso-octane.

The mechanism of conversion of n-octane may possibly be as follows:

$$C_8H_{18} \rightarrow 2C_4H_9, \qquad C_2H_5 + H \rightarrow C_2H_4 + H_2,$$
$$C_4H_9 \rightarrow C_2H_4 + C_2H_5, \qquad C_2H_5 \rightarrow CH_2 + CH_3,$$
$$C_2H_5 \rightarrow C_2H_3 + H_2, \qquad C_2H_5 + CH_3 \rightarrow C_2H_4 + CH_4.$$
$$C_2H_3 \rightarrow C_2H_2 + H,$$

The following processes are also possible:

$$C_2H_4 \rightarrow 2CH_2, \qquad C_3H_7 \rightarrow C_2H_4 + CH_3,$$
$$CH_2 + CH_3 \rightarrow C_2H_3 + H_2, \qquad C_2H_5 + CH_2 \rightarrow C_3H_4 + CH_3,$$
$$C_2H_3 \rightarrow C_2H_2 + H, \qquad CH_3 + H \rightarrow CH_4,$$
$$CH_3 + H \rightarrow CH_2 + H_2, \qquad R + R + M \rightarrow R_2 + M.$$
$$C_8H_{18} \rightarrow C_3H_7 + C_5H_{11},$$

TABLE 1. Characteristics of Pyrolysis of Gasoline and of Individual Hydrocarbons in a Hydrogen Plasma. Mean Specific Arc Energy 1.2 kWh/m³

Characteristic	n-Hexane	n-Heptane	n-Octane	Iso-octane	Gasoline
Maximum C_2H_2 concentration, vol.%	5.3	5.9	6.3	6.3	5.3
Maximum C_2H_4 concentration, vol.%	11.3	12.0	13.1	9.0	8.8
Maximum $(C_2H_2 + C_2H_4)$ concentration, vol.%	16.2	16.1	18.0	14.2	13.2
Maximum corrected concentration of unsaturated compounds, vol.%	58.1	59.7—65.4	62.3	54.9	56.0

The conversion scheme of iso-octane (2,2,4-trimethylpentane) can also be written as follows:

$$C_8H_{18} \rightarrow H_3C - C - CH_2 + CH - CH_3$$
$$\underset{CH_3 \; CH_3}{\bigwedge} \quad \underset{CH_3}{|}$$

$$\begin{matrix} H_3C \\ {}_{}^{} \\ H_3C \end{matrix} \hspace{-6pt} > \hspace{-4pt} CH \rightarrow C_2H_4 + CH_3,$$

$$\begin{matrix} H_3C \\ H_3C - C - CH_2 \rightarrow \\ H_3C \end{matrix} \quad \begin{matrix} H_3C \\ {}^{} \\ H_3C \end{matrix} \hspace{-6pt} > \hspace{-4pt} C - CH_3 + CH_2$$

$$\begin{matrix} H_3C \\ H_3C - C \rightarrow \\ H_3C \end{matrix} \quad \begin{matrix} H_3C \\ {}^{} \\ H_3C \end{matrix} \hspace{-6pt} > \hspace{-4pt} C + CH_3,$$

$$H_3C - C - CH_3 \rightarrow C_2H_3 + CH_3,$$

$$C_2H_3 \rightarrow C_2H_2 + H, \qquad CH_3 + CH_2 \rightarrow C_2H_3 + H_2,$$
$$CH_3 + H \rightarrow CH_4, \qquad R + R + M \rightarrow R_2 + M.$$

The following processes are also possible:

$$C_2H_3 + H \rightarrow C_2H_4,$$
$$C_2H_3 + H \rightarrow C_2H_2 + H_2.$$

The number of C—C bonds broken here is about half as much again as in the pyrolysis of n-octane. This agrees with the relative large methane concentration in the pyrolysis gas of iso-octane and with the reduced expenditure of energy required to pyrolyse n-octane.

Table 1 gives the quantitative relations between the maximum reaction-product concentrations from gasoline pyrolysis and from pyrolysis of individual hydrocarbons, when the specific arc energy averages 1.2 kWh/m³ H_2.

A very important fact is that, both for individual hydrocarbons and gasoline, when the pyrolysis is conducted so that the hydrocarbons do not reach the discharge, the overall conversion of hydrocarbons is close to

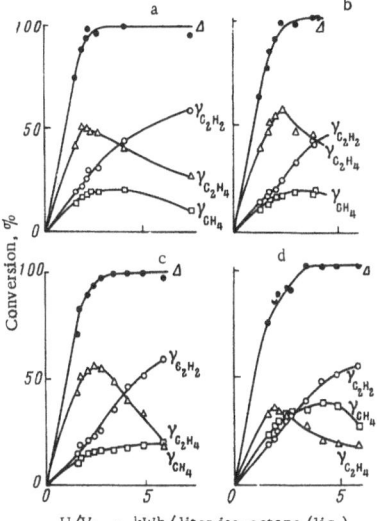

Fig. 5. Overall conversion of hydrocarbons (Δ) and conversion of hydrocarbons to C_2H_2, C_2H_4, and CH_4 (γ) in a hydrogen plasma, plotted versus specific pyrolysis energy U/V_{hyd}. Pyrolysis of: a) n-hexane; b) n-heptane; c) n-octane; d) iso-octane.

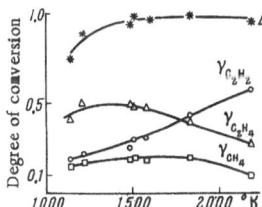

Fig. 6. Overall conversion of n-hexane (Δ) and degrees of conversion to C_2H_2, C_2H_4, and CH_4 (γ), plotted versus temperature.

unity and there is practically no soot or other solid products.

This experimental fact is illustrated in Fig. 5, in which the conversion of n-hexane, n-heptane, n-octane, and iso-octane, calculated from the mass balance of hydrocarbons, is plotted versus the specific pyrolysis energy U/V_{hyd}. It will be seen that the overall conversion Δ of all the hydrocarbons to gaseous products reaches unity for specific energy 3 kWh per liter hydrocarbon (liq.). For conversion of hydrocarbons to acetylene ($\gamma_{C_2H_2}$), the increase is characteristically proportional to the specific energy. The conversion to ethylene, $\gamma_{C_2H_2}$, always has a maximum at specific energies of about 2-2.5 kWh/liter hydrocarbon (liq.), reaching 50-60% for normal hydrocarbons and 35% for iso-octane. For all the normal hydrocarbons, the conversion to methane, γ_{CH_4}, reaches a maximum at values of U/V_{hyd} equal to about 4-5 kWh/liter hydrocarbon (liq.). In the case of iso-octane, owing to the reduced pyrolysis to ethylene, the value is higher (37%).

In Fig. 6 the degree of conversion of n-hexane is plotted versus the temperature of the final gas. It will be seen that at about 1500°K, the overall conversion gets close to 100%. The degree of conversion to acetylene increases with temperature, while the curves for ethylene and methane pass through a maximum. However, these maxima are not very marked, and the temperatures corresponding to the maximum values of $\gamma_{C_2H_2}$ and γ_{CH_4} can only be estimated very roughly.

If as ordinate we plot, not the experimental degrees of conversion, but the ratios of the standard changes of Gibbs free energy $(\Delta Z_T^\circ)_i/(\Delta Z_T^\circ)_\Sigma$, which can be regarded nominally as the relative probabilities of the various branches of the reaction in comparison with the observed sum of all the reactions (Fig. 7), then we get a picture which to some extent agrees with the experiments.

In fact, when T > 1500°K, the curves corresponding to the probabilities of formation of C_2H_4, CH_4, and C_2H_2 are arranged in a similar way to the experimental values. However, the probability of decomposition of n-hexane to its elements, $(\Delta Z_T^\circ)_i/(\Delta Z_T^\circ)_\Sigma$, predominates so much over the other branches of the reaction that the practical absence of soot in the reaction products is due to the relative slowness of the first reaction.

Finally, we would note that blank experiments, with specific energies greater than the maximum energy in the pyrolysis experiments, revealed that there is practically no interaction between the hydrogen plasma and the graphite of the lower electrode; this agrees with the failure of attempts to synthesize acetylene by adding graphite dust to a hydrogen plasma [4].

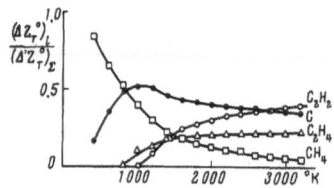

Fig. 7. Ratio $(\Delta Z_T^\circ)_i/(\Delta Z_T^\circ)_\Sigma$ versus temperature. $(\Delta Z_T^\circ)_i$ is the change of Gibbs free energy for conversion of n-hexane to C_2H_2, C_2H_4, CH_4, or C; $(\Delta Z_T^\circ)_\Sigma$ is the sum of these changes.

SUMMARY

1. The authors studied the pyrolysis of vapor of low-octane gasoline and the individual hydrocarbons $n\text{-}C_6H_{14}$, $n\text{-}C_7H_{16}$, $n\text{-}C_8H_{18}$, and iso-C_8H_{18} in a hydrogen plasma obtained from a high-voltage dc arc at pressures of about 1.2-2 atm and with an equivalent temperature of up to 5000°C. The process was regulated so that no hydrocarbons reached the discharge zone.

2. It was found that the qualitative composition of the final products does not vary much with the nature of the initial material. Both gasoline (a mixture of 76 individual substances) and the individual hydrocarbons studied pyrolyse mainly to acetylene, ethylene, and methane. Thus gasoline behaves like an individual saturated hydrocarbon.

3. The overall conversion of gasoline to gaseous compounds is close to 100%, and there is very little soot or other solid products.

4. The degree of conversion of gasoline to unsaturated compounds (acetylene + ethylene) is 80%, while the maximum concentration of unsaturated compounds is 17.5 vol.% (of which up to 9 vol.% is C_2H_2), which constitutes 56 vol.% of the pyrolysis gases.

5. For gasoline, the minimum energy consumption α, which corresponds to the maximum total concentration of unsaturated compounds, is close to 4 kWh/m^3 $C_2H_2 + C_2H_4$; this means that 54% of the energy of the hydrogen arc is expended on the formation of acetylene and ethylene.

6. In the pyrolysis of vapors of normal hydrocarbons, the results are similar in all respects to the results of gasoline pyrolysis; there is some increase in the mass characteristics as the number of carbon atoms in the original molecule rises.

7. There is an anomaly in the pyrolysis of iso-octane: in contrast to the case of normal octane, the maximum ethylene concentration and the degree of conversion to ethylene are lower than the corresponding values for methane.

8. Branching of the carbon chain does not affect the formation of acetylene: this suggests that the mechanism of acetylene formation is different from that of ethylene formation.

9. Pyrolysis in a hydrogen plasma is more efficient than other pyrolytic processes in respect of the total concentration of acetylene and ethylene, which reaches 65% of the cracking gas by volume.

LITERATURE CITED

1. Wulf, R. G., US Patent 2236555 (1941).
2. Chem. Trade (review). J. Chem. Engng., No. 3684: 142 (1958).
3. Gurvich, L. V., G. A. Khachkuruzov, B. A. Medvedev, et al., Thermodynamic Properties of Individual Substances, Izd. Akad. Nauk SSSR, Moscow (1962).
4. Leutner, H. W., and C. S. Stokes, Ind. Engng. Chem., 53 : 341 (1961).
5. Cagas, F., M. Staud, and A. Lasarew, Erdöl u. Kohle, 12: 818 (1959).
6. Il'in, D. T., and E. N. Eremin, Zh. Fiz. Khim., 36:1560 (1962).
7. Rossini, F. D., K. S. Pitzer, et al., Selected Values of Physical and Thermodynamic Properties of Hydrocarbons and Related Compounds, Carnegie Press, Pittsburgh (1953).
8. Anderson, J. E., and L. K. Case, Ind. Engng. Chem., Process Design Develop., 1 : 161 (1962).

9.   Eremin, E. N., and D. T. Il'in, Russian Patent No.152236 (1951); Byull. Izobr., No. 10: 92 (1963).
10.   Il'in, D. T., and E. N. Eremin, Vestn. Moskovsk. Univ., No. 2: 48 (1962).
11.   Cagas, F., M. Staud, and A. Lasarew, Chem. Prumysl., 9:185 (1959).

HOW THE ELECTROCRACKING OF METHANE IS AFFECTED
BY ADDING PROPANE AND BUTANE

D. T. Il'in and E. N. Eremin

As shown by Kobozev and Shneerson [1] and later by Gordon [2] and Germain and Vaniscotte [3], in the thermal cracking of methane with added ethane, decomposition of the ethane induces cracking of the methane.

We have attempted to study how the addition of small amounts of higher hydrocarbons affects the electrocracking of methane in a dc arc at pressures of about 1.5 atm.

As additive we used a propane—butane mixture (ratio of propane to butane, 4 : 1 by volume), which we added to the methane in the (average) proportions 2.82, 3.69, and 7.32% by volume.

The apparatus was described in [4]. To this we added only a system for feeding in the propane—butane additive: this consisted of a cylinder of propane—butane mixture connected, via a tap and a dosing flowmeter fitted with a mercury manometer, to the methane inlet. The general experimental method was also described in [4].

The gas additive was analyzed by means of a TsIATIM (Central Research Institute for Aviation Fuel and Oil) apparatus.

The mixture of cracking methane and additive was analyzed with a VTI (All-Union Fuel Technology Institute) apparatus. The unreacted part of the additive was heated with hydrogen in a quartz loop filled with powdered copper oxide (350°C). In these conditions the residual propane and butane were converted to carbon dioxide, as was the carbon monoxide contained in the gas. But since the percentage content of CO_2 remained practically the same as in the experiments without additive, it followed that all the propane and butane reacted at the high temperatures of the arc. In these conditions, the content of higher compounds in the decomposition products of the additive could not be large; owing to the low additive concentration, it could not in fact exceed a fraction of a percent. This conclusion agrees with literature data [5] on the pyrolysis of propane at 1100°C. Our method of finding the overall conversion Δ_{CH_4} of the methane involved measuring the carbon balance. We assumed that all the propane—butane additive was converted to acetylene, ethylene, and its elements. Note that with this method, the value found for the total methane conversion can only be smaller than the true value, because probably some of the additive was converted to methane. We also tried a somewhat different method of calculating Δ_{CH_4}, but it yielded practically the same results (see Fig. 2).

The figures plot the results of experiments on the influence of propane—butane additive in the electrocracking of methane. The process characteristics are as follows: volume percentage content of acetylene; total volume percentage content of acetylene and ethylene; energy expenditure (α, kWh/m^3 C_2H_2); and degree of overall methane conversion (Δ_{CH_4}). These were plotted versus the specific energy (Figs. 1 and 2) and versus the volume percentage content of additive V_1/V (Fig. 3).*

---

*V_1 is the volume of additive.

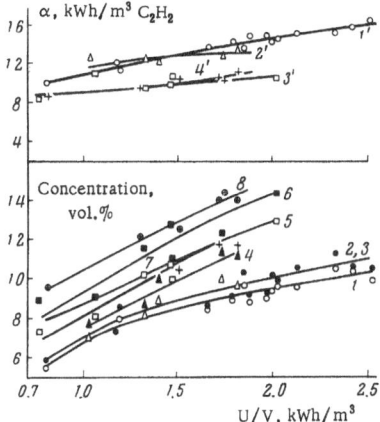

Fig. 1. Concentrations of C_2H_2 and $(C_2H_2 + C_2H_4)$ and energy expenditure α, plotted versus specific energy U/V in experiments on joint electrocracking of methane and propane—butane additive. 1,2,1') No additive; 3,4,2') 2.82 vol.% additive; 5,6,3') 3.69 vol.%; 7,8,4') 7.32 vol.%.

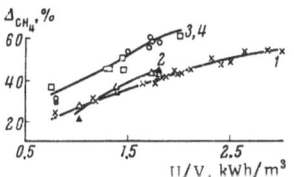

Fig. 2. Total conversion of methane Δ_{CH_4} versus specific energy U/V in experiments on joint electrocracking of methane and propane—butane additive. 1) No additive; 2) 2.82 vol.% additive; 3) 3.69 vol.%; 4) 7.32 vol.%.

Returning to the graphs (Figs. 1 and 2) of the process characteristics versus the specific energy U/V, we see that they rise with increasing specific energy: the best process characteristics correspond to the greatest additive content.

For a specific energy of, for example, 2.00 kWh/m^3, and with 3.69% additive, the greatest improvement in the process characteristics is expressed in the following increase of the yields: C_2H_2, from 9.5 to 13.0 vol.%; $(C_2H_2 + C_2H_4)$, from 10.2 to 14.4 vol.%; degree of overall methane conversion, from 42 to 62%; reduction in energy consumption from 15.0 to 10.8 kWh/m^3 C_2H_2, or from 14.3 kWh/m^3 $C_2H_2 + C_2H_4$ to 9.73 kWh/m^3 $C_2H_2 + C_2H_4$.

As shown by Fig. 1, the energy expenditures in the experiments without additive (curve 1') practically coincide with those for minimum additive (curve 2'), and the energy expenditures in the experiments with 3.69% additive (curve 3') practically coincide with those for 7.32% additive (curve 4').

The graphs of overall methane conversion versus specific energy display a similar pattern (see Fig. 2).

These relations between energy expenditure and overall methane conversion become more intelligible when we consider Fig. 3, which plots the mass and energy characteristics of electrocracking, for a specific energy of 1.82 kWh/m^3, versus the volume percentage content of propane—butane additive.

Figure 3 shows that between 0 and 2.5 vol.% of additive, and also above 4 vol.%, all the process characteristics vary relatively slowly, whereas between 2.5 and 4% additive they improve suddenly. Thus, in this region the C_2H_2 content rose from 9.7 to 12.3 vol.%, and that of $C_2H_2 + C_2H_4$ from 10.9 to 13.8 vol.%, while the energy expenditure α fell from 13.4 to 10.3 kWh per m^3 C_2H_2. The overall methane conversion increased from 43 to 60%; this will be especially important for the argument below.

It is thus clear that, for a given specific energy, addition of up to 2.5% propane and butane has relatively little effect on methane conversion; but greater additions lead to fairly marked intensification of the interaction between the components of the reaction mixture. Up to 4% additive, the increase in the mass characteristics is proportional to the additive concentration in the original mixture, as shown by the dashed lines in Fig. 3. Further increase in the amount of additive then leads to less marked changes in the process characteristics. We can thus consider that the excess additive above 4% concentration is cracked independently. On the whole, we can apparently speak of a process of joint electrocracking of methane and additive, which has the result that some additional methane is drawn into the reaction. There is a limiting additive concentration (approximately 4 vol.% of propane—butane mixture in the original gas for the given specific energy), corresponding to the end of the joint process. As shown by Figs. 1 and 2, this values is approximately the same for other specific energies also. We can interpret the nature of the process as follows.

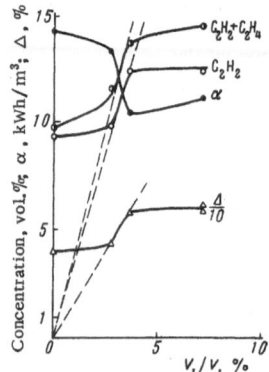

Fig. 3. Process characteristics of electrocracking of methane versus volume concentration of propane–butane additive, V_1/V. $U/V = 1.82$ kWh/m³.

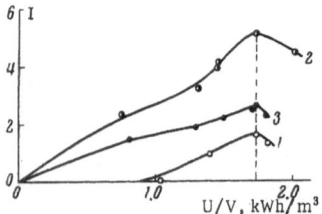

Fig. 4. Induction coefficient I versus specific energy U/V, in experiments on joint electrocracking of methane and added propane–butane mixture. Additive concentrations (vol.%): 1) 2.82; 2) 3.69; 3) 7.32.

Since propane and butane decompose more easily than methane, the additive is a source of free radicals, which draw additional methane into the reaction. The decomposition products of propane and butane intensify the cracking of methane: this is similar to the joint pyrolysis of methane and ethane observed by Kobozev and Shneerson [1].

These considerations are supported by the marked rise (from 0.315 to 0.430 m³/kWh) which occurs in the mean energetic efficiency $(k_1 + k_2)$ of the arc when the content of propane–butane additive rises from 2.82 to 3.69%.

We must emphasize that the energetic efficiency was calculated from the overall methane conversion, and corresponds to the amount of methane converted per kilowatt-hour of energy when $U/V \to 0$.

From the increase in the overall methane conversion we can easily find the induction coefficient I of the process:

$$I = \frac{(\Delta_{I, II, III} - \Delta_0) V}{V_1} ,$$

where $\Delta_{I,II,III}$ and Δ_0 are the overall methane conversions in experiments with and without additive, respectively, V is the rate of consumption of methane, and V_1 is the rate of consumption of propane–butane additive.

Figure 4 plots the calculated induction coefficients. We see that in all three series of experiments, I at first increases with the specific energy, and then passes through a maximum at $U/V = 1.73$ kWh/m³.

The reader will notice that the maxima of I occur at approximately the same value of U/V. For 2.82% additive (curve 1), the maximum induction coefficient is about 1.6 mole CH_4 per mole additive; for 3.69% additive, the figure is 5.2 mole CH_4 per mole additive, and for 7.32% additive, 2.6 mole CH_4 per mole additive. At U/V values close to maximum I, the additive decomposes into the most active particles with large stocks of energy. The decrease in I which accompanies further increase in U/V is probably due to decomposition of the additive into less active substances which do not promote induction.

Because of the high induction coefficients, together with the shapes of the $(I, U/V)$ graphs, we can conclude that the interaction of methane with the products of additive decomposition in the arc is a chain reaction.

In the absence of additive, the supposed mechanism of methane conversion in the arc is as follows:

$$CH_4 \to CH_3 + H,$$
$$CH_3 \to CH_2 + H,$$
$$CH_4 + H \to CH_3 + H_2,$$
$$CH_3 + CH_2 \to C_2H_3 + H_2,$$

$$C_2H_3 \to C_2H_2 + H,$$
$$CH_4 + CH_2 \to 2CH_3,$$
$$CH_4 + CH_3 \to C_2H_5 + H_2,$$
$$C_2H_5 \to CH_2 + CH_3.$$

Other possible processes are

$$CH_2 \rightarrow CH + H,$$
$$CH + CH_4 \rightarrow C_2H_4 + H,$$
$$C_2H_4 \rightarrow 2CH_2,$$
$$CH + CH_4 \rightarrow C_2H_3 + H_2,$$
$$CH + CH_3 \rightarrow C_2H_2 + H_2,$$
$$CH + CH_2 \rightarrow C_2H_2 + H,$$

$$CH_3 + CH_3 \rightarrow C_2H_6,$$
$$C_2H_6 \rightarrow 2CH_3,$$
$$C_2H_6 \rightarrow C_2H_5 + H,$$
$$nCH_2 \rightarrow C_nH_{2n},$$
$$R + R + M \rightarrow R_2 + M.$$

The mechanism by which the additive decomposes and draws additional methane into the reaction can be represented as follows:

$$C_3H_8 \rightarrow C_2H_5 + CH_3,$$
$$C_4H_{10} \rightarrow 2C_2H_5,$$
$$C_2H_5 \rightarrow CH_2 + CH_3,$$
$$CH_3 + CH_4 \rightarrow C_2H_6 + H,$$
$$C_2H_6 \rightarrow 2CH_3,$$
$$CH_3 \rightarrow CH_2 + H,$$

$$CH_4 + H \rightarrow CH_3 + H_2,$$
$$CH_3 + CH_2 \rightarrow C_2H_3 + H_2,$$
$$C_2H_3 \rightarrow C_2H_2 + H,$$
$$CH_4 + C_2H_5 \rightarrow CH_3 + C_2H_6,$$
$$C_2H_5 \rightarrow CH_3 + CH_2,$$
$$C_2H_5 \rightarrow C_2H_3 + H_2.$$

Other possible processes are

$$CH_2 \rightarrow CH + H,$$
$$CH_4 + CH \rightarrow C_2H_3 + H_2,$$
$$C_2H_3 \rightarrow C_2H_2 + H,$$
$$CH_3 + H \rightarrow CH_2 + H_2,$$

$$CH_3 + CH \rightarrow C_2H_2 + H_2,$$
$$CH_2 + CH \rightarrow C_2H_2 + H,$$
$$C_2H_6 \rightarrow C_2H_4 + H_2,$$
$$C_2H_4 \rightarrow C_2H_2 + H_2.$$

A comparison of our data with the results of Kobozev and Shneerson shows that a propane–butane additive has greater inductive effectiveness in the electric-arc process than an ethane additive has in pyrolysis. In the former of these two cases, the coefficient of induction reaches 5.2 mole CH_4 per mole additive, or about 1.6 mole CH_4 per g-atom of carbon in the additive; in the second case (pyrolysis), it reaches 2 mole CH_4 per mole additive, or 1 mole CH_4 per g-atom carbon.

The ability to induce methane conversion is evidently a property possessed to some extent in common by all the saturated hydrocarbons. As the chain length of the added hydrocarbon increases further, the maximum induction coefficient expressed in mole CH_4 per mole additive will apparently increase, while that expressed in mole CH_4 per g-atom C in additive will tend to a constant limit.

SUMMARY

1. The authors studied the joint cracking of methane and propane–butane additive ($C_3H_8 : C_4H_{10} = 4:1$); the proportions of additive to original gas were 2.82, 3.69, and 7.32 vol.%, and the dc arc was at a pressure of about 1.5 atm.

2. It was found that the additives improve the characteristics of the electrocracking of methane.

3. All the characteristics of the process vary with the additive concentration: the variation is relatively small for additive concentrations of 0-2.5 or more than 4% by volume, but between 2.5 and 4% the characteristics improve suddenly because additional methane is drawn into the reaction.

4. The drawing of additional methane into the reaction is apparently due to the inductive action of the additive's decomposition products.

5. The induction coefficients I of the process were found. For each series of experiments, the graph of I versus specific energy U/V passes through a maximum at $U/V = 1.73$ kWh/m^3. For an additive concentration of 3.69%, $I_{max} = 5.2$ moles CH_4 per mole additive.

6. Owing to the high induction coefficient and the shape of the $(I, U/V)$ graphs, we conclude that the interaction of methane with the products of decomposition of the additive in the arc is a chain reaction.

LITERATURE CITED

1.   Kobozev, N. I., and A. L. Shneerson, Dokl. Akad. Nauk SSSR, 33: 217 (1941).
2.   Gordon, A. S., J. Am. Chem. Soc., 70: 395 (1948).
3.   Germain, J. E., and C. Vaniscotte, Bull. Soc. Chim. France, No. 6: 692 (1957); No. 3: 319 (1958); No. 7: 964 (1958).
4.   Il'in, D. T., and E. N. Eremin, Zh. Prikl. Khim., 35: 2064 (1962).
5.   Morina, I. N., Chemical Processing of Petroleum Hydrocarbons. Izd. Akad. Nauk SSSR, Moscow (1956), p. 98.

HOW THE DIMENSIONS OF THE REACTION CHANNEL
AND DISCHARGE CHAMBER AFFECT THE ELECTROCRACKING
OF METHANE

D. T. Il'in and E. N. Eremin

There is practically no literature information on how the dimensions of the reaction zone and the geometrical shape of the reaction path affect the cracking of methane in an arc in dynamic conditions. Nevertheless, the geometry of the reaction space is both qualitatively and quantitatively important in the process, for it is this factor which determines the reaction time τ, and with given τ but differently shaped discharge chambers and reaction channels it may affect the electrical and gas-dynamic conditions of the discharge, and hence also the process as a whole. Movement of gas in the reactor governs the shape of the arc (elongates it and stabilizes it), effects contact of the gas with the active zone, and (what is very important) carries solid products (soot) away from the reaction zone.

Our work was concerned mainly with the influence of the cross-sectional area of the reaction channel S, and of the volume of the discharge chamber V_{ch}, on the electrocracking of methane.

We used an apparatus which was previously described in [1]: the initial gas contained about 95% of methane.

The pressure in the reactor was 1-1.5 atm. The arc current was 2-15 A dc, and its voltage 500-3500 V. The power expended was 25 kW. The reactor consisted of a steel chamber with a tangential gas inlet and vertical electrodes. The gas left the chamber via the lower, grounded electrode, in which there was a channel forming a continuation of the reaction zone.

We used lower electrodes with reaction channels of diameters 4, 5, 5.5, 6, 10, and 15 mm, and length 170 mm, and also with diameter 10 mm and length 100 mm. However, in the conditions of our experiments [1], it was practically impossible to use the electrodes with 4-mm and 15-mm channels, because when the specific energy exceeded 1-1.3 kWh/m^3,* in the former case the pressure in the discharge chamber became too great, and in the latter case the sides of the channel became too rapidly covered with soot and carbon deposits. Incidentally, it was therefore very difficult to refer the process characteristics for very low acetylene yields to any particular channel cross section.

We used discharge chambers with volumes of 400, 200, 100, and 50 cm^3. The last two of these were found to be unsuitable, owing to electrical breakdown via soot deposited on the chamber walls.

We tested lower and upper electrodes with channels and holes of three configurations (Fig. 1A, B); the shape of the chamber remained practically constant. In respect to electrocracking, we found that channels shaped as symmetrical nozzles (b) or double nozzles (c) behaved in our conditions in about the same way as a

---

* All volumes reduced to NTP — Translator.

D. T. IL'IN AND E. N. EREMIN

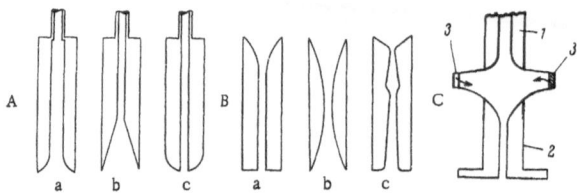

Fig. 1. Shapes of electrodes and discharge chamber. A) High-voltage electrode; B) grounded electrode; C) discharge chamber (1, high-voltage electrode; 2, grounded electrode; 3, tangential gas inlets).

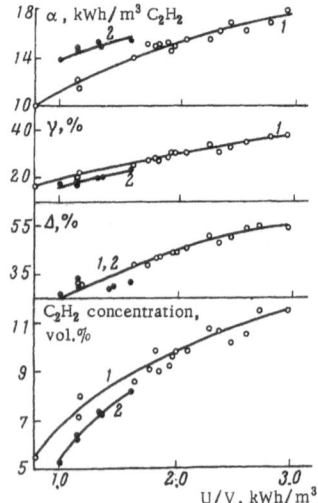

Fig. 2. C_2H_2 concentration, overall methane conversion Δ, degree of conversion of methane to acetylene γ, and energy expenditure α, plotted versus specific energy U/V. Lengths of reaction channel in mm: 1) 100; 2) 170.

cylindrical channel with a funnel-shaped inlet (a) of equal volume. We therefore continued to use channels with the simplest construction (a).

For the upper (high-voltage) electrode, we found that the main factor was the hole diameter; for diameters of 8, 12, and 15 mm, the arc scarcely entered the hole and attached itself to a limited number of points on the electrode end; this promoted the formation of carbon dendrites which reduced the length of the discharge gap, and in some cases permitted rapid burnup of the electrode sides. Only with holes of diameter 18-26 mm did the discharge get pulled into the hole for a distance of 20-40 mm, while the cathode spot rotated uniformly around the circumference of the cavity and we observed practically no dendrite formation. In view of this, we chose a high-voltage electrode with a cavity diameter of 26 mm.

The end of the upper electrode was shaped as shown in Fig. 1. For a given hole diameter, electrodes of shapes a, b, and c behaved almost identically. In the work we used an electrode with a funnel-shaped inlet hole.

The experimental method and the main methods of processing the results were described in a previous paper [1]. In comparing the process characteristics, we used an additional variable — the cross-sectional area S of the reaction channel, which is known to depend (for a homogeneous reaction zone) on the reaction time τ, the reaction volume V_r, the reaction-zone length L, and the volumetric and linear velocities, V and V_l; the formula for these relations is

$$S = \frac{V}{V_l} = \frac{V\tau}{L} = \frac{V_r}{L}. \tag{1}$$

Figures 2-6 plot the results of experiments to elucidate how the dimensions of the channel in the lower electrode and of the discharge chamber affect the electrocracking of methane.

Figure 2 plots the process characteristics versus the specific energy: these were measured with lower electrodes with a cylindrical channel of 10 mm diameter and 100 mm length (curves 1), or 170 mm length (curves 2); the discharge chambers were of the same volume (400 cm³) and shape. It will be seen that better results are

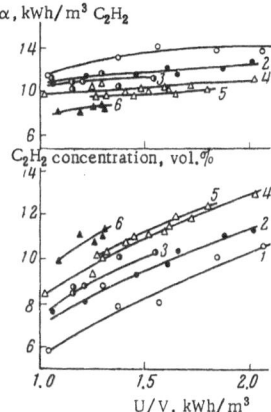

Fig. 3. C_2H_2 concentration and energy consumption α, plotted versus specific energy U/V.
1) Reaction channel diameter $d = 10$ mm, $V_{ch} = V_0 = 400$ cm^3; 2) $d = 10$ mm, $V_{ch} = V_0/2$; 3) $d = 6$ mm, $V_{ch} = V_0$; 4) $d = 5.5$ mm, $V_{ch} = V_0/2$; 5) $d = 5$ mm, $V_{ch} = V_0/2$; 6) $d = 5.5$ mm, $V_{ch} = V_0/2$. Heating to 500°C.

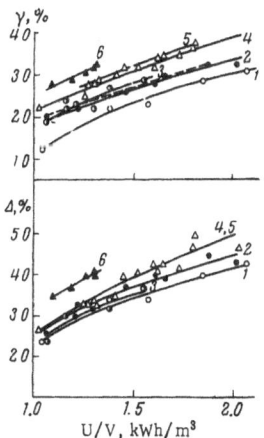

Fig. 4. Overall conversion Δ and degree of conversion of methane to acetylene γ, plotted versus specific energy U/V. For meanings of numbers see Fig. 3.

obtained with shorter channels. In particular, with specific energy 1.6 kWh/m^3, the C_2H_2 content rose from 8 to 8.7 vol.%, while the energy consumption fell from 15.6 to 13.7 kWh/m^3 C_2H_2. It is interesting that the changes were mainly in the characteristics associated with the accumulation of acetylene. At the same time there was little change in the overall methane conversion Δ_{CH_4}; this confirmed that, in these two series of experiments, the mean energetic discharge efficiencies *
$k_1 + k_2 = (V/U)\ln(1 - \Delta_{CH_4})$ were almost the same (0.278 and 0.270 m^3/kWh). Thus a decrease in reaction channel length leads mainly to improved preservation of the acetylene formed, and has little effect on processes occurring in the arc itself. Further decrease in the channel length was limited by the maximum penetration of the arc into the channel.

We shall show that changes in the cross-sectional area of the channel and in the volume of the discharge chamber exert more influence on the results than do reductions in the length of the reaction channel: they lead not only to improvements in acetylene preservation, but also to better methane-conversion kinetics.

The energy expenditure α, acetylene concentration in vol.%, degree of overall conversion Δ_{CH_4}, and degree of conversion to acetylene $\gamma_{C_2H_2}$ (Fig. 4), obtained when working with reaction channels of various cross-sectional areas and with two chambers of different volumes, are plotted in Figs. 3 and 4 versus the specific energy. It will be seen that all the curves show a typical [2, 3] rise with the specific energy. The highest process characteristics occur with the smallest channel cross section and the smaller chamber. For example, if the specific energy is 2.04 kWh/m^3 and the channel diameter is 10 mm, when we go from a chamber of volume V_0 (400 cm^3) (curves 1) to one of volume $V_0/2$ (curves 2), the mass characteristics increase as follows: acetylene concentration from 10.3 to 11.5 vol.%; overall methane conversion from 41.2 to 44.5%; degree of conversion of methane to acetylene from 31.0 to 34.5%. The energy consumption decreases from 14.3 to 12.6 kWh/m^3.

In these conditions, the energetic efficiency of the arc rose from 0.265 to 0.296 m^3/kWh.

With the same specific energy, and with a chamber of volume $V_0/2$, when the channel diameter was reduced from 10 mm (curves 2) to 5.5 mm (curves 4), the mass

---

*For a more detailed discussion of the energy efficiency, see [2].

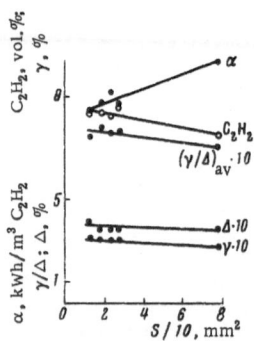

Fig. 5. Characteristics of electrocracking of methane, plotted versus cross-sectional area of reaction channel S. $U/V = 1.36$ kWh/m^3; $V_{ch} = V_0/2$.

characteristics also improved: the acetylene concentration increased from 11.5 to 13.0 vol.%, the overall conversion from 44.5 to 49%, the degree of conversion to acetylene from 33.5 to 40.4%. The energy consumption fell from 12.6 to 11.0 kWh/m^3 C$_2$H$_2$, i.e., by about 15%. The energetic efficiency of the discharge thus increased from 0.296 to 0.318 m^3/kWh. Thus, in these conditions, a twofold decrease in the chamber volume corresponds to a threefold decrease in the cross-sectional area of the reaction channel.

Figure 5 plots the main process characteristics for all the channel diameters used (for $U/V = 1.36$ kWh/m^3, $V = V_0/2$). All the above characteristics, with the addition of the ratio $\gamma_{C_2H_2}/\Delta_{CH_4}$, are plotted versus S.

It will be seen that, as S increases, in the range studied, the process characteristics, like the C$_2$H$_2$ content, Δ_{CH_4}, $\gamma_{C_2H_2}$, and $\gamma_{C_2H_2}/\Delta_{CH_4}$, decrease, while α increases.

Since in the ranges studied all the curves are practically linear, we can characterize the relations between the process characteristics and the channel cross section by means of the differences between the mean values obtained in experiments with the channels of least (4 mm) and greatest (10 mm) diameters. As shown by Fig. 5, these differences are as follows: acetylene concentration, 1.3 vol.% (9.3 and 8 vol.%); Δ_{CH_4}, 4% (38 and 34%); $\gamma_{C_2H_2}$, 5% (31 and 26%); α, 2.3 kWh/m^3 C$_2$H$_2$ (9.3 and 11.6 kWh/m^3 C$_2$H$_2$). The value of $\gamma_{C_2H_2}/\Delta_{CH_4}$ increases from 75 to 85%.

It is easy to verify that the graph of the process characteristics versus S is similar for higher specific energies.

The increase of $\gamma_{C_2H_2}/\Delta_{CH_4}$ is noteworthy: it provides evidence that the decomposition of methane and newly formed acetylene to their elements is largely suppressed by reduction of the reaction volume and hence of the reaction time. In many cases this ratio reaches 80-90%, as against 60-70% obtained with a large chamber and a wide channel.

Together with the increase in acetylene concentration, another important factor is the reduction in energy consumption achieved by optimum choice of chamber and channel — in our experiments, this averaged 30%. For example, by using the maximum specific energy for these experiments, i.e., 2.03 kWh/m^3, with a 5.5 mm diameter channel and a chamber of volume $V_0/2$ (curves 4 in Figs. 3 and 4), the characteristics improved as follows over the results of using a 10-mm diameter channel and a chamber of volume V_0 (curves 1 in Figs. 3 and 4); the acetylene concentration increased from 10.3 to 13 vol.%; the overall methane conversion increased from 41 to 49%; the degree of conversion to acetylene increased from 31 to 40%; and the energy consumption decreased from 14.3 to 11.0 kWh/m^3 C$_2$H$_2$. The fraction of the methane which was converted to acetylene, $(\gamma_{C_2H_2}/\Delta_{CH_4})_{av}$, increased from 67 to 82%. In these experiments, the energetic efficiency of the discharge increased on average from 0.265 to 0.318 m^3/kWh.

In Fig. 6, the ratios $\gamma_{C_2H_2}/\Delta_{CH_4}$, γ_m/γ_n, $(k_1 + k_2)_m/(k_1 + k_2)_n$, and α_n/α_m (where m and n are the numbers of the curves in Figs. 3 and 4) are plotted versus the specific energy. We notice that these ratios are independent of the specific energy in the range studied, and that $(k_1 + k_2)_m/(k_1 + k_2)_n$ practically coincides with Δ_m/Δ_n, and γ_m/γ_n with α_n/α_m.

The increases in the overall conversion and in the degree of conversion of methane to acetylene can be qualitatively attributed to the following causes:

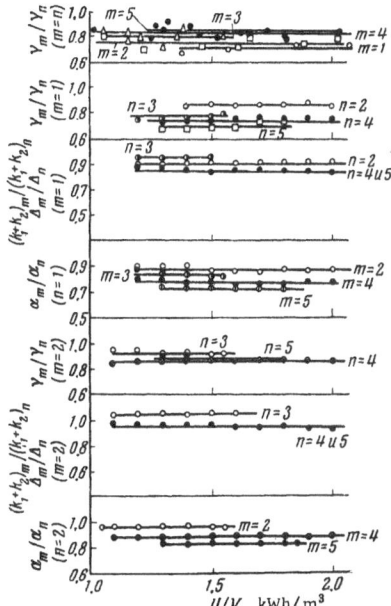

Fig. 6. Ratios of process characteristics and energetic efficiency ($k_1 + k_2$), plotted versus specific energy U/V. m, n are the numbers of the curves in Figs. 3 and 4.

1. Improvement in the gas—dynamic characteristics, for which the reaction conditions approach the optimum.

2. Decrease in unfavorable gas circulation in the reaction zone, which increases the time of residence of the reaction products in this zone and hence causes them to decompose further into their elements.

3. Increase in the linear speed of the gas, which promotes entrainment of soot out of the reaction zone.

4. Improvement in the conditions of contact of the gas with the reaction zone.

5. Change in lateral temperature profile and mass distribution in the reaction channel, leading to a reduction in the cold ring-shaped gap between the "pinch" of the arc and the sides of the channel. (In this connection, reduction of S and V_{ch} has the same effect as preheating of the methane (see Figs. 3 and 4, curves 6).*

The results can also be treated by an approach based on the equations of Vasil'ev, Kobozev, and Eremin [2, 3] for the overall degree of conversion of methane to acetylene,

$$\Delta_{CH_4} = 1 - e^{-(k_1 + k_2) U/V}, \qquad (2)$$

for the degree of conversion of methane to acetylene,

$$\gamma_{C_2H_2} = \frac{k_1}{k_1 + k_2 - k_3} [1 - e^{-(k_1 + k_2 - k_3) U/V}] e^{-k_3 U/V} \qquad (3)$$

and for the energy consumption,

$$\alpha = \frac{2(k_1 + k_2 - k_3) \frac{U}{V} e^{+k_3 \frac{U}{V}}}{k_1 (1 - e^{-(k_1 + k_2 - k_3) U/V})} \qquad (4)$$

These equations were derived from the kinetic scheme

$$2CH_4 \underset{k_2}{\overset{k_1}{\lessgtr}} \begin{array}{c} C_2H_2 \\ \downarrow k_3 \\ C + H_2, \end{array}$$

where k_1, k_2, and k_3 are the velocity constants of the corresponding first-order reactions, independent of the power U, and having the dimensions of m^3/kWh.

---

* The influence of preheating of the methane on the characteristics of electrocracking is discussed in detail in [4-7].

Expand (2) in a series of powers of U/V:

$$\Delta_{CH_4} = (k_1 + k_2)\,\frac{U}{V} - \frac{(k_1 + k_2)^2(U/V)^2}{1 \cdot 2} + \frac{(k_1 + k_3)^3(U/V)^3}{1 \cdot 2 \cdot 3} + \ldots \tag{5}$$

and, for small values of U/V, reject terms after the first. Then we get [4] the following expression for Δ_{CH_4}:

$$\Delta_{CH_4} = (k_1 + k_2)\,U/V, \tag{6}$$

where $k_1 + k_2$, as already stated, is called the energetic efficiency of the discharge, and denotes the limiting quantity of methane (in m^3) which (for small U/V) is converted by one kilowatt-hour of arc energy.

Similarly, by expanding (3) and (4) in series, provided that we can neglect the decomposition of acetylene into its elements, we get

$$\gamma_{C_2H_2} = k_1\,\frac{U}{V}, \tag{7}$$

$$\alpha = \frac{2}{k_1}. \tag{7a}$$

By (6) and (7), at low specific energies Δ_{CH_4} and $\gamma_{C_2H_2}$ are proportional to U/V, whence

$$\gamma_{C_2H_2}/\Delta_{CH_4} = \frac{k_1}{k_1 + k_2} = \text{const.} \tag{8}$$

This is what is actually found in the experiments (see Fig. 6).

Furthermore, if we assume that the total result of electrocracking is additively composed of the results of cracking in the discharge chamber and in the channel, and if we use the idea of a mean linear velocity $\overline{V}_l$ for two series of experiments with different reaction-zone cross sections S, then, by (1), (6), and (7) we get

$$\frac{\overline{S}_m}{\overline{S}_n} = \frac{(k_1 + k_2)_m}{(k_1 + k_2)_n}\,\frac{\Delta_n}{\Delta_m}\,\frac{(U/\overline{V}_l)_m}{(U/\overline{V}_l)_n} = \frac{(k_1)_m}{(k_1)_n}\,\frac{\gamma_n}{\gamma_m}\,\frac{(U/\overline{V}_l)_m}{(U/\overline{V}_l)_n}. \tag{9}$$

If we compare the process characteristics obtained in two series of experiments with the same specific energy, by (6) and (7) we get

$$\frac{(k_1 + k_2)_m}{(k_1 + k_2)_n}\bigg/\frac{\Delta_m}{\Delta_n} = \frac{(k_1)_m}{(k_1)_n}\bigg/\frac{\gamma_m}{\gamma_n} = 1, \tag{10}$$

whence

$$\Delta_m/\Delta_n = \frac{(k_1 + k_2)_m}{(k_1 + k_2)_n} = \text{const} \tag{11}$$

and

$$(k_1)_m/(k_1)_n = \gamma_m/\gamma_n = \text{const.} \tag{12}$$

From Fig. 6 we see that (11) and (12) are confirmed by experiment.

It follows from (7a) that, on the above assumptions, the energy consumption α should be constant for each series of experiments. But on returning to Fig. 3, we see that α does vary with U/V, although only slightly.

However, if we take the ratio of the consumptions for any two series of experiments, then the relation

$$\frac{\alpha_n}{\alpha_m} = \frac{(k_1)_m}{(k_1)_n} = \text{const}, \tag{13}$$

which follows from (7a), will also be true experimentally.

It follows that $\alpha_n/\alpha_m = \gamma_m/\gamma_n$, which is approximately satisfied by the results in Fig. 6.

Remembering that the reaction-zone volume V_r is included in (2) via the kinetic constants k_1 and k_2, we get the following relation:

$$\bar{V}_r = \frac{\Delta_{CH_4}}{\dfrac{U}{V}(k_1^* + k_2^*)} = \frac{\gamma_{C_2H_2}}{\dfrac{U}{V}k_1^*}, \tag{14}$$

where k_1^* and k_2^* are kinetic constants without allowance for V_r.

This means that, in terms of the final result, decrease of the reaction-zone volume is formally equivalent to increase of specific energy U/V. This agrees with the purely qualitative theory given above.

We can get a relation between the reaction-zone dimensions (particularly between the cross-sectional areas of this zone — if we compare the results of two series of experiments with the same, not too great specific energy) by beginning from (9), which, by (1), (11), and (12), can be transformed to

$$\bar{S}_m/\bar{S}_n = (U/\bar{V}_l)_m/(U/\bar{V}_l)_n \tag{15}$$

and

$$(\bar{V}_r)_m/(\bar{V}_r)_n = (U\bar{\tau})_m/(U\bar{\tau})_n. \tag{16}$$

It must be emphasized that these relations are valid only insofar as the appropriate approximations and assumptions are true of any particular case. They are certainly true in general at low specific energies, where the process characteristics depend linearly on the specific energy. In other cases the relations are only approximate, and indicate the main tendencies in the behavior of the process characteristics; for exact calculations we must make use of (2)-(4), as we did, for example, in finding the energetic efficiency of the discharge.

SUMMARY

1. The authors study how the dimensions of the reaction channel in the lower electrode and of the discharge chamber affect the characteristics of the electrocracking of methane in a dc arc under a pressure of about 1.5-2 atm.

2. Reduction in the reaction-channel length leads mainly to improved preservation of acetylene, and has little effect on the processes taking place in the arc itself. At the same time, reduction in the cross-sectional area of the channel and in the discharge-chamber volume exerts an influence on all the process characteristics, including the overall methane conversion.

3. The proportion of cracking to form acetylene, $\gamma_{C_2H_2}/\Delta_{CH_4}$, obtained when working with a minimum-volume reaction zone, is greater than that obtained when working with a maximum-volume reaction zone (80-90% as against 60-70%).

4. The energetic efficiency of the discharge ($k_1 + k_2$) increases with the channel diameter and with the chamber volume.

5. These results are explained by the suppression of decomposition of acetylene due to reduction in the reaction volume, and also by the changes which occur in the temperatures and gas dynamics of the process.

LITERATURE CITED

1. Il'in, D. T., and E. N. Eremin, Khim. Prom., No. 6:408 (1962).
2. Vasil'ev, S. S., N. I. Kobozev, and E. N. Eremin, Zh. Fiz. Khim., 7:619 (1936).
3. Eremin, E. N., Khim. Prom., No. 2:14 (1958).
4. Eremin, E. N., Vestn. Moskovsk. Univ., No. 3:54 (1961).
5. Vasil'ev, S. S., Vestn. Moskovsk. Univ., No. 12:63 (1947).

6.   Il'in, D. T., and E. N. Eremin, Zh. Prikl. Khim., 35:2064 (1962).
7.   Il'in, D. T., and E. N. Eremin, Zh. Prikl. Khim., 35:2496 (1962).

KINETICS AND MECHANISM OF ACETYLENE CONVERSION
IN A GLOW DISCHARGE AT LOW PRESSURES

E. N. Borisova and E. N. Eremin

Much attention has been devoted to methane conversion in electric discharges since the middle of the 19th century, and many reports have been published. However, despite this interest and the commercial application of the electrocracking of methane to acetylene, very little is known regarding the mechanism of methane conversion. It would appear that the process is extremely complex and includes a number of stages with formation of various stable intermediate products (C_2H_6, C_2H_4, C_2H_2). Each stage may itself consist of several elementary reactions. Free atoms and radicals must participate in the conversion mechanism, because they are known to be present in discharges. The authors of [1-3] detected CH, C_2, and H particles spectroscopically, and Willey [4] and Letort [5] also reported the presence of CH_3 and CH_2 particles.

From the wealth of experimental data, various reaction mechanisms have been postulated. The mechanism suggested by Kassel [6] for pyrolysis of methane has been extended to its electrocracking. The mechanism may be represented schematically as follows:

$$CH_4 \rightleftarrows C_2H_6 \rightleftarrows C_2H_4 \rightleftarrows C_2H_2 \rightleftarrows C + H_2.$$

Many investigators [7-12] have postulated that acetylene formation in a discharge takes place in stages via ethane and ethylene as intermediate products.

For a complete analysis of methane conversion in a discharge we must therefore know the kinetic laws governing the decomposition of all the stable intermediate products. The present authors made preliminary experiments [13, 14] in connection with a comparative study of the decomposition kinetics of methane and acetylene in a glow discharge at 28-36 mm Hg; these showed that formation of acetylene exclusively or predominantly via ethylene is improbable.

The aim of the present work is to study the kinetics and mechanism of the electrocracking of methane to acetylene by investigating the conversion kinetics of all the hydrocarbons which might participate in the reaction, beginning with acetylene as the end link in the chain and ending with methane itself.

To obtain conditions enabling us to accumulate considerable amounts of the supposedly less stable products (ethane and ethylene) as well as the most stable product (acetylene), we selected an ac glow discharge at a low pressure (1-2 mm Hg) and a continuous-flow procedure. It was assumed that these were the most favorable conditions for effecting and detecting the sequential reactions of methane conversion.

The experiments were performed in a glass vacuum apparatus (Fig. 1). The main part of the apparatus was a quartz reactor of diameter 2 cm and length 90 cm (1). The aluminum-tipped steel electrodes (2) (an upper high-voltage electrode and an earthed lower electrode) were cooled by tap water. The pressure in the discharge zone of the apparatus was measured by a McLeod gauge (3) and a vacuum gauge (5).

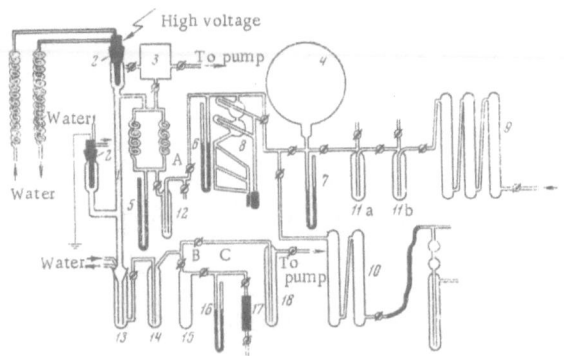

Fig. 1. Apparatus for conversion of gaseous hydrocarbons in a low-pressure
glow discharge.

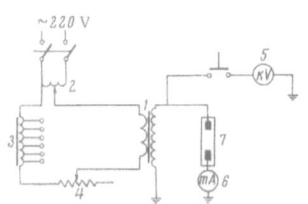

Fig. 2. Electrical circuit of the apparatus.
1) Transformer; 2) autotransformer; 3)
choke; 4) rheostat; 5) kilovoltmeter; 6)
milliammeter; 7) reactor.

The feed gas was contained in a 10-liter cylinder (4), and its pressure was measured by manometers (6) and (7); the flow rate through the reactor was controlled by a tap (A) and rheometer (8). The volume of the cylinder was measured accurately and the amount of gas consumed calculated from the difference between the pressures. The feed gases were purified before entering the cylinder by passing through columns with silica gel and caustic soda (9, 10) and traps (11a, 11b). Methane was subjected to further purification by removal of heavier hydrocarbons in a liquid-nitrogen trap (12).* The products were condensed at the discharge tube outlet in traps (13, 14) at the temperature of liquid nitrogen. After the end of an experiment the reaction products were refrozen in trap (15). The volume of this trap, together with those of manometer (16) and taps B and C were accurately measured. We thus determined the amount of condensed gaseous reaction products. Gas analysis was performed in a KhT-2M chromathermograph; the analysis samples were taken by a syringe from a rubber hose (17). We determined the following components: air, methane, ethane, ethylene, propane, propylene, and acetylene. Since the activities of the platinum elements varied, we calibrated the chromathermograph before each experiment. Ahead of the pump, the gas passed through a dust trap (18). Figure 2 shows the electrical circuit of the apparatus.

The kinetics of methane conversion were studied at a discharge-tube pressure 0.7-1.7 mm Hg and gas flow rate 2.5-15 liters/h†; the current strength in the discharge was kept at 32 mA, which was the minimum strength in the experiments with other hydrocarbons. The feed gas was commercial acetylene, purified by passing through concentrated H_2SO_4.

Figure 3 plots $\Delta_{C_2H_2}$, the degree of total conversion of acetylene (the ratio of the amount of reacted acetylene to the amount passed through the discharge), versus the specific energy U/V (where U is the discharge power in W, and V is the volumetric flow rate of the gas in liters/h). It will be seen that the reaction is intense;

---

* Trap 12 was not used in the experiments on acetylene, ethylene, and ethane.
† Here and below the gas volumes are converted to NTP.

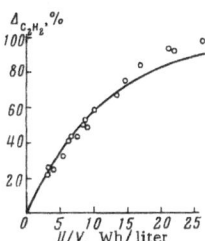

Fig. 3. Degree of total conversion of acetylene ($\Delta_{C_2H_2}$) versus the specific energy (U/V); k = 0.086 liter/Wh. The experimental data are superimposed on the theoretical curve.

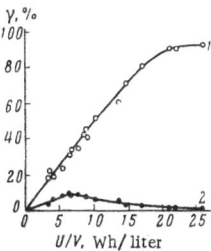

Fig. 4. The degree of acetylene conversion (γ) to the solid products (1) and gaseous acetylene derivatives (2), plotted versus the specific energy (U/V).

with U/V = 8.5 Wh/liter, half the acetylene is already decomposed, and at 20 Wh/liter conversion is almost complete.

From reports on discharge reactions and the thermal decomposition of acetylene, it is known that acetylene conversion is described by a first- or second-order reaction, depending on the conditions. We calculated the velocity constants for a first-order reaction by the equation of Vasil'ev, Kobozev, and Eremin [15]:

$$k_1 = -\frac{1}{U/V} \ln\left(1 - \Delta_{C_2H_2}\right), \text{ liters/Wh.}$$

The values of the constants remained practically unchanged, with the exception of a slight increase at high specific energies. From their mean value (0.0855 liter/Wh) we plotted the degree of acetylene transformation versus the specific energy (cf. the solid curve in Fig. 3). It will be seen that the curve agrees with the experimental data up to nearly 80% transformation.

The main conversion product of acetylene in a discharge consists of solid brown polymers, found on the walls of the tube and traps after the experiment. The gaseous products included ethane, ethylene (γ, the degree of acetylene conversion to these, was less than 0.1%), traces of propane and propylene, and gaseous derivatives of acetylene. A separate analysis of the latter showed the presence of more than ten different compounds, but only three (methylacetylene, vinylacetylene, and diacetylene) were identified. The percentage content of the sum of the gaseous acetylene derivatives was found by subtracting from 100% the percentage content of the other components of the condensate (C_2H_2, C_2H_6, C_2H_4, C_3H_8, and C_3H_6). For calculating the degrees of conversion of acetylene to its gaseous derivatives (Fig. 4, curve 2), we assumed that they have the mean composition C_4. It will be seen that curve 2 has the characteristic shape for an intermediate product, reaching a maximum (9%) at about 6 Wh/liter. The graph of the degree of acetylene conversion to solid products (Fig. 4, curve 1) is a curve with saturation at higher U/V. At relatively low specific energies (less than 18 Wh/liter) the experimental values lie on a straight line passing through the origin.

Attempts to use a higher current strength (75 mA) did not give data for the complete curve because such experiments were accompanied by an increase in the discharge power and the specific energies. It was therefore impossible to obtain data near the origin, and at high U/V practically all the acetylene was converted. From the data for two experiments at 75 mA, the velocity constant of acetylene decomposition displayed little increase.

It was found that acetylene decomposition is already intense at low pressures and discharge powers. It is of interest to know how it is formed and retained in large amounts, for example during methane cracking under much more severe conditions. To verify the possibility of its stabilization by the cracking products of methane, consisting mainly of hydrogen in addition to acetylene, we performed experiments with 1:1, 1:1.8, and 1:5 mixtures of acetylene and hydrogen under the same discharge conditions as for pure acetylene. The velocity constants of acetylene conversion were calculated from the equation

$$k_1 = -\frac{1}{U/V_{C_2H_2}} \ln\left(1 - \Delta_{C_2H_2}\right), \text{ liters } C_2H_2/Wh,$$

where $V_{C_2H_2}$ is the partial velocity of acetylene in the mixture in liters/h.

The mean values of the constants were 0.065, 0.050, and 0.031 liter/Wh, respectively, i.e., they decreased with the dilution of acetylene. It is noteworthy that, in the case of conversion of pure acetylene mixed with once or twice its volume of hydrogen, k_1 remains constant or virtually constant with increasing U/V. In the latter case, when the ratio $C_2H_2 : H_2 = 1 : 5$, the velocity constant of the first-order reaction decreases with increasing specific energy. We also calculated the velocity constants for the second-order reaction by the equation

$$k_2 = -\frac{1}{U/V_{C_2H_2}} \frac{[C_2H_2]}{[C_2H_2]_0 \left([C_2H_2]_0 - [C_2H_2]\right)}.$$

It was found that for the first two dilutions k_2 increases with the specific energy, but at maximum dilution k_2 is more constant than k_1 (Table 1).

A comparison of the cracking products of the acetylene—hydrogen mixture and of acetylene itself showed that the results differ markedly only in the last stage. For example, whereas the conversion of acetylene to ethane and ethylene was less than 0.4% in the first two stages, with a 1 : 6 ratio of acetylene and hydrogen the amount of the hydrogenation products of acetylene was greatly increased. It is of interest that the predominant component in the hydrogenation products was not ethylene, as would be expected, but ethane. The ethane content in the condensate reached 15% ($\gamma_{C_2H_6} \approx 5\%$), while that of ethylene was only 3% ($\gamma_{C_2H_4} \approx 1\%$).

A similar increase was observed in the amount of propane, propylene, and gaseous derivatives of C_2H_2 ($\gamma \approx 10\%$).

From these experimental data we can infer that conversion of acetylene (and acetylene—hydrogen mixtures) in a low-pressure glow discharge is described by a first-order equation. The energy velocity constants of the first-order reaction decrease with dilution of acetylene by hydrogen. The gaseous derivatives are intermediate products in the acetylene-conversion reaction. Hydrogenation of acetylene is very small and increases appreciably only with marked dilution by hydrogen. The presence of methylacetylene and other C_3 hydrocarbons in the reaction products indicates that the process is fairly complex.

Any mechanism of acetylene conversion must naturally give a satisfactory explanation of the experimental data. Thus, the fact that acetylene conversion is a first-order reaction eliminates simple molecular polymerization, which occurs in low-temperature thermal reactions [15, 16].

In a low-pressure glow discharge, characterized by the presence of high electron and low molecular temperatures, the conditions favor the formation of considerable amounts of free radicals and atoms. In the case of acetylene, the electron-excited molecules can decompose in two ways [17]:

TABLE 1. Variation of k_1 and k_2 (in liters C_2H_2/Wh) with Specific Energy
U/V (in Wh/liter)

$C_2H_2 : H_2 = 1 : 1$			$C_2H_2 : H_2 = 1 : 1.8$			$C_2H_2 : H_2 = 1 : 5$		
U/V	k_1	k_2	U/V	k_1	k_2	U/V	k_1	k_2
7.4	0.061	0.078	8.5	0.047	0.057	14.4	0.039	0.053
9.3	0.061	0.082	9.5	0.047	0.059	16.0	0.038	0.053
10.3	0.068	0.097	16.1	0.048	0.073	19.7	0.031	0.042
11.3	0.062	0.089	20.5	0.051	0.089	33.2	0.029	0.042
12.9	0.062	0.089	21.4	0.051	0.092	36.1	0.031	0.056
20.4	0.074	0.173	33.0	0.056	0.164	54.8	0.028	0.069
						62.9	0.025	0.060

$$C_2H_2 \rightarrow C_2H + H, \qquad \Delta H^\circ = 113 - 115 \ \text{kcal/mole}, \tag{1}$$

$$C_2H_2 \rightarrow 2CH, \qquad \Delta H^\circ = 230 \pm 2 \ \text{kcal/mole}. \tag{2}$$

Both reactions are endothermic, but (1) is more probable and may very well be the first stage of the process. (Note that under these conditions we observe, for example, dissociation of hydrogen to atoms, $\Delta H^\circ = 103$ kcal per mole.) The hydrogen atoms may then react with the acetylene as follows:

$$C_2H_2 + H \rightarrow C_2H_3, \tag{3}$$

$$C_2H_2 + H \rightarrow C_2H + H_2. \tag{4}$$

In our experiments, (4) is apparently more important than (3), leading to hydrogenation of acetylene, because hydrogenation products of ethylene and ethane were found in negligible amounts. C_2H radicals formed by (1) and (4) react further with acetylene, ultimately leading to its polymerization:

$$C_2H + C_2H_2 \rightarrow C_4H_3, \tag{5}$$

$$C_4H_3 + C_2H_2 \rightarrow C_6H_5, \tag{6}$$

etc.

They may also to some extent participate in the formation of diacetylene, which is found in the reaction products:

$$2C_2H \rightarrow C_4H_2.$$

Also possible is a further decomposition under the action of electron impacts:

$$C_2H \rightarrow C_2 + H.$$

C_2 particles were found spectroscopically when acetylene reacted in a discharge.

Selecting the most probable processes for the kinetic calculations, we get the following list of reactions:

$$C_2H_2 \xrightarrow{k_1} C_2H + H, \tag{1}$$

$$C_2H_2 + H \xrightarrow{k_4} C_2H + H_2, \tag{4}$$

$$C_2H + C_2H_2 \xrightarrow{k_5} C_4H_3, \tag{5}$$

$$H \xrightarrow{k_7} \frac{1}{2} H_2. \tag{7}$$

Here we are assuming that recombination of atomic hydrogen takes place mainly at the walls, because the triple collisions required for recombination of H in the reaction space are improbable at low pressures.

Processing of the results by the stationary concentrations method gives the following equation for the overall velocity of acetylene conversion:

$$- \frac{d\,[C_2H_2]}{dt} = 2k_1\,[C_2H_2]\left(1 + \frac{k_4\,[C_2H_2]}{k_4\,[C_2H_2] + k_7}\right) \approx 4k_1\,[C_2H_2].$$

If we assume that k_7 is much less than the product $k_4[C_2H_2]$, we get a first-order equation, in agreement with experiment.

However, we cannot altogether exclude a molecular conversion mechanism. It is known that the glow-discharge zone in the tube is not uniform. In addition to dissociation by electron impact, the molecule may be activated without undergoing complete decomposition. Furthermore, the molecules may undergo thermal activation, i.e., we may have conditions akin to low-temperature conditions, which favor molecular polymerization of acetylene:

$$2C_2H_2 \rightarrow C_4H_4,$$

$$C_4H_4 + C_2H_2 \rightarrow C_6H_6, \text{ etc.}$$

What effect can dilution of acetylene with hydrogen have on the mechanism?

It is known that in a glow-discharge hydrogen dissociates into atoms to a considerable extent. It might be throught that the presence of atomic hydrogen should increase the role of (4), thus leading eventually to an increase in the decomposition velocity of acetylene. However, it follows from our experiments that the reverse is the true picture, evidently because the following reaction occurs readily and rapidly in the presence of high H concentrations:

$$C_2H + H \rightarrow C_2H_2; \tag{8}$$

this leads to regeneration of acetylene. This reaction competes with (5), which leads to the end products. In other words, recombination of the hydrogen atoms to acetylene takes place via (8) and (4). Similar catalytic recombination has been observed by many authors [16] who have studied thermal reactions of acetylene with atoms of hydrogen and deuterium.

Thus, according to this mechanism, the conversion of acetylene diluted with hydrogen can be described by such reactions as

$$C_2H_2 \overset{k}{\rightarrow} C_2H + H, \tag{1}$$

$$C_2H_2 + H \overset{k_4}{\rightarrow} C_2H + H_2, \tag{4}$$

$$C_2H + H \overset{k_8}{\rightarrow} C_2H_2, \tag{8}$$

$$C_2H + C_2H_2 \overset{k_5}{\rightarrow} C_4H_3, \tag{5}$$

$$H_2 \overset{k_9}{\rightarrow} 2H, \tag{9}$$

$$H \overset{k_7}{\rightarrow} \frac{1}{2} H_2. \tag{7}$$

Processing by the stationary concentrations method gives the following equation for the conversion velocity of acetylene:

$$-\frac{d[C_2H_2]}{dt} = 4k_5 [C_2H_2]^2 \left[\frac{2k_1k_4 [C_2H_2] + k_1k_7 + k_4k_9 [H_2]}{2k_8k_9 [H_2] + 2k_4k_5 [C_2H_2]^2 + k_5k_7 [C_2H_2]} \right].$$

At high [H_2], the main role in both the numerator and denominator is played by terms containing the hydrogen concentration, and all the other terms can be neglected. We then get a second-order equation:

$$-\frac{d[C_2H_2]}{dt} \simeq \frac{2k_5k_4}{k_8} [C_2H_2]^2.$$

SUMMARY

1. We have studied the kinetics of acetylene conversion in a glow discharge at low pressures (0.7-1.7 mm Hg).

2. It has been established that pure acetylene is converted by a first-order reaction to form mainly solid brown polymers and hydrogen. The degree of conversion to gaseous acetylene derivatives (methylacetylene, vinylacetylene, diacetylene, etc.) reaches 9% of the initial acetylene. The degree of conversion to ethane and ethylene is less than 0.1%.

3. Dilution with hydrogen retards acetylene decomposition, i.e., it reduces the energy velocity constant. At high dilutions the overall reaction changes to a second-order reaction with a considerable increase in the yields of ethane, ethylene, propane, and propylene.

4. Mechanisms are postulated for the conversion of acetylene itself and of acetylene diluted with hydrogen.

LITERATURE CITED

1. Peters, K., and O. H. Wagner, Z. Phys. Chem., A153:161 (1931).
2. Kolyubin, A. A., Izv. Akad. Nauk SSSR, ser. fiz., 22:753 (1958).
3. Starodubtsev, S. V., Sh. A. Abalyaev, L. Ya. Alimova, and Yu. B. Sokolova, Izv. Akad. Nauk UzbSSR, ser. fiz.-mat., No. 5:68 (1962).
4. Willey, E. J., Trans. Faraday Soc., 30:230 (1934).
5. Letort, M., and X. Duval, Compt. Rend., 219:452 (1944).
6. Kassel, L. S., J. Am. Chem. Soc., 54:3949 (1932).
7. Peters, L., and A. Pranschke, Brennstoff-Chem., No. 11:239 (1930).
8. Saint-Auney, R. V., Chim. Ind., 29:1011 (1933).
9. Amemiya, T., J. Soc. Chem. Ind. Japan, 41:371 (1938).
10. Starodubtsev, S. V., Sh. A. Abalyaev, F. Bakhramov, Sh. Ziyatdinov, and L. G. Keitlin, Izv. Akad. Nauk UzbSSR, ser. fiz.-mat., No. 5:60 (1962).
11. Eremin, E. N., Dissertation, MGU (1951).
12. Weeler, T. S., Fuel, No. 10:175 (1930).
13. Borisova, E. N., and E. N. Eremin, Zh. Fiz. Khim., 36:1261 (1962).
14. Borisova, E. N., and E. N. Eremin, Zh. Fiz. Khim., 36:2234 (1962).
15. Vasil'ev, S. S., N. I. Kobozev, and E. N. Eremin, Zh. Fiz. Khim., 7:619 (1936); Khim. Tv. Topliva, No. 8:70 (1937).
16. Steacie, E. W. R., Atomic and Free Radical Reactions, 2nd ed. Reinhold, New York (1954).
17. Vedeneev, V. I., L. V. Gurvich, V. N. Kondrat'ev, V. A. Medvedev, and E. L. Frankevich, Bond Energies, Ionization Potentials and Electron Affinities. St. Martin's Press, New York (1966).

KINETICS AND MECHANISM OF ETHYLENE CONVERSION
IN A GLOW DISCHARGE AT LOW PRESSURES

E. N. Borisova and E. N. Eremin

It is known [1-18] that under the effect of electric discharges ethylene can take part in a wide range of reactions: conversion to acetylene, methane, carbon, and hydrogen, hydrogenation to ethane, polymerization to liquid and solid hydrocarbons, etc. The direction taken by ethylene conversion depends on the type of discharge and its power, and on the experimental conditions — particularly the pressure in the discharge zone. It was therefore of interest to study the behavior of ethylene in a glow discharge at low pressures, under the same conditions in which the conversion of acetylene was studied.

For this work we used commercial ethylene (purity 99.9%). The experiments were performed in a glow discharge at discharge-tube pressures of 0.7-1.7 mm Hg. The flow rate of the gas varied from 1 to 20 liters/h (under normal conditions), and the power of the discharge from 26 to 588 W. We performed six series of experiments, at current strengths 32, 75, 125, 175, 225, and 350 mA. The apparatus and the experimental method are described in [19].

Chromatographic analysis revealed that, in addition to unreacted ethylene, the condensate contained several gaseous reaction products (acetylene, ethane, propane, and propylene). Figure 1 plots ethylene conversion (the ratio of the amount of ethylene converted to a particular product to the total amount of ethylene passed through the discharge) versus the specific energy U/V (Wh/liter). It will be seen that the main conversion product of ethylene is acetylene. The curves plotting $\gamma_{C_2H_2}$ versus the specific energy pass through maxima. *

In all the series of experiments, the $\gamma_{C_2H_2}$ maxima are reached approximately with U/V = 30 Wh/liter, and the height of the maximum depends on the current strength in the discharge, i.e., on its power. For example, $\gamma_{C_2H_2}$ max at 32 mA reaches only 24%, but at 350 mA more than 34%. After the maximum has been reached, the fall of the curves is retarded with decreasing current strength in the discharge.

The dehydrogenation of ethylene to acetylene is accompanied by its hydrogenation to ethane by the hydrogen formed in the reaction. The curves plotting $\gamma_{C_2H_6}$ versus the specific energy have broad indistinct maxima, which shift slightly toward high U/V with decreasing current strength in the discharge. In all the series of experiments the maxima hardly differed at all from one another and were less than 6%.

The amounts of propane and propylene formed were small. The degrees of transformation of ethylene to C_3 hydrocarbons also pass through maxima which vary with U/V; $\gamma_{C_3H_8}$ reaches a maximum (about 2.5 %) very rapidly at approximately 10-20 Wh liter and then falls slowly. In all the series of experiments, $\gamma_{C_3H_6}$ is less than 1.5%.

---

*In the case of the highest current strengths (350 mA) the maximum is apparently located at unattainable gas flow rates.

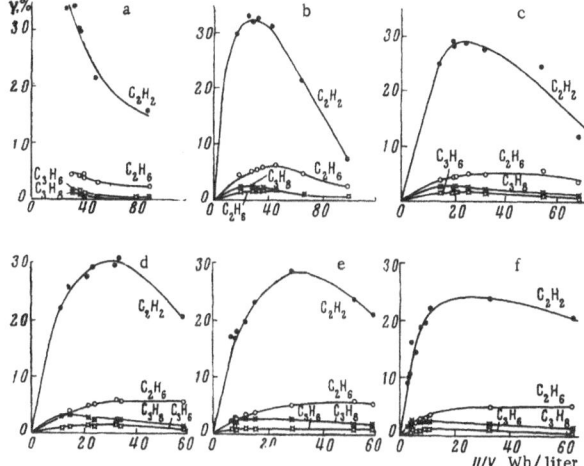

Fig. 1. Ethylene conversion to the end products (γ) versus the specific energy U/V. Current strength (in mA): a) 350; b) 225; c) 175; d) 125; e) 75; f) 32.

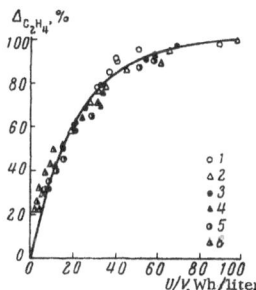

Fig. 2. Degree of total conversion of ethylene ($\Delta_{C_2H_4}$) versus the specific energy U/V. Current strength (in mA): 1) 350; 2) 225; 3) 175; 4) 125; 5) 75; 6) 32.

During the experiments, solid reaction products were deposited on the walls of the discharge tube and the traps; they consisted of carbon and a brown tarry substance. Conversion of ethylene to the solid products reached very high values, ranging from 10% at very low U/V to 80% at high specific energies, when ethylene decomposition is virtually complete (96-98%).

Figure 2 plots the degree of total conversion of ethylene $\Delta_{C_2H_4}$ (the ratio of the amount of decomposed ethylene to the total amount passed through the discharge) versus U/V for all the current strengths. It will be seen that the experimental points lie very close to a single curve with a mean first-order velocity constant equal to 0.046 liter/Wh (from the equation of Vasil'ev, Kobozev, and Eremin [18]). Therefore, to a first approximation the degree of total conversion may be represented (for all the current strengths) as a single function of the specific energy. However, whereas the constants hardly vary throughout the whole range of change of the specific energy at 350, 225, 175, and 125 mA in each series of experiments, the mean values of the constants have a tendency to decrease with falling current strength in the discharge: 0.052, 0.044, 0.044, and 0.043 liter/Wh, respectively. It is also noteworthy that in the case of the smaller current strengths the constants vary and depend on the experiment's specific energy. At 75 mA the velocity constant of the first-order reaction decreases from 0.058 to 0.040 with increasing specific energy, and at 32 mA it decreases to practically a third (from 0.094 to 0.035 liter/Wh). It is possible that the decrease in the power of the discharge is accompanied by a change in the mechanism, so that the process is no longer described by a first-order equation. In fact, in the case of the experiments with current strength 32 mA, the values of the velocity constants seemed more correct when calculated for a second-order reaction. At current strength 75 mA, the velocity constants of the first-order reaction decreased with the specific energy, whereas those of the second-order reaction increased.

TABLE 1. Variation of End Gas Composition with the Specific Gravity

Amt. of ethylene fed, V, liter/h	Electric characteristics		End gas composition, vol.%						
	U, W	U/V, Wh/liter	H_2	CH_4	C_2H_6	C_2H_4	C_2H_2	C_3H_8	C_3H_6
15.7	386	24.6	50.5	2.4	3.5	21.7	20.0	1.2	1.3
8.93	323	36.2	63.3	2.6	4.1	12.2	16.5	1.3	1.2
5.41	284	52.4	77.3	3.3	3.1	4.7	10.4	—	—

TABLE 2. Velocity Constants as Functions of the Current Strength

Current strength, mA	$a_2 = k_2$	$a_3 = k_1' + k_3 + k_9$	$a_4 = k_4$	k_1'	k_3	k_9
225	0.018	0.044	0.022	0.005	0.028	0.011
175	0.015	0.044	0.030	0.004	0.030	0.010
125	0.013	0.043	0.030	0.004	0.030	0.009

Since this experimental procedure did not enable us to monitor the conversion of ethylene to methane and hydrogen, we made an attempt to collect the end gases after the backing pump without prefreezing in liquid-nitrogen traps. Owing to gas losses in the pump, we could not obtain entirely accurate data; however, we were able to follow the relative concentration of the products in the final cracking gas. We performed several experiments at current strength 225 mA and pressure 1-2 mm Hg; the results are given in Table 1.

The methane concentration is approximately the same as that of ethane; therefore, γ_{CH_4}, the degree of transformation of ethylene to methane, should be about half the value of $\gamma_{C_2H_6}$, i.e., it may reach 2-3%.

From an analysis of the curves plotting the concentration of the ethylene conversion products (to be more exact, the corresponding degrees of conversion — cf. Fig. 1), we postulated the following kinetic scheme for the reactions:

$$
\begin{array}{c}
\overset{k_9}{\overbrace{\qquad\qquad}} \\
C_2H_4 \overset{k_3}{\rightarrow} C_2H_2 \rightarrow \text{solid products}, \; H_2, \; CH_4, \; C_3H_8, \; C_3H_6. \\
k_2' \searrow \quad \nearrow k_2 \\
C_2H_6
\end{array}
$$

According to this scheme, ethylene decomposes in three parallel directions: conversion to acetylene (k_3), hydrogenation to ethane (k_2'), and conversion to other products, which may be methane, propane, propylene, and solid substances (k_9). As the end reaction products in this scheme, we took (in addition to the solid substances and hydrogen) all those compounds which, in the given case, may participate to a lesser degree in acetylene formation, i.e., methane, propane, and propylene. Therefore we shall not examine the formation and decomposition of the individual hydrocarbons and disregard the undoubtedly small amounts of ethane, ethylene, and acetylene which may be formed from them. Furthermore, we shall disregard the reverse reactions (hydrogenation of acetylene and dehydrogenation of ethane), because, as shown by the experiments on acetylene and ethane, their degrees of conversion to ethylene are small and the amount of ethylene formed from them can be neglected in comparison with its initial amount.

Bearing in mind that the decomposition of C_2H_2, C_2H_4, and C_2H_6 was found experimentally to be of the first order, we can write a system of differential equations for the rates of formation of these hydrocarbons:

$$- \frac{d\,[C_2H_4]}{dt} = (k'_2 + k_3 + k_9)\,[C_2H_4],$$

$$\frac{d\,[C_2H_2]}{dt} = k_3\,[C_2H_4] - k_4\,C[_2H_2],$$

$$\frac{d\,[C_2H_6]}{dt} = k'_2\,[C_2H_4] - k_2\,[C_2H_6].$$

Solution of these gives the following expressions for the degrees of conversion:

$$1 - \Delta_{C_2H_4} = e^{-a_3 U/V}, \tag{1}$$

$$\gamma_{C_2H_2} = k_3 \left(\frac{1}{a_4 - a_3} e^{-a_3 U/V} + \frac{1}{a_3 - a_4} e^{-a_4 U/V} \right), \tag{2}$$

$$\gamma_{C_2H_6} = k'_2 \left(\frac{1}{a_3 - a_2} e^{-a_2 U/V} + \frac{1}{a_2 - a_3} e^{-a_3 U/V} \right). \tag{3}$$

Here, $a_3 = k_3 + k_9 + k'_2$, $a_4 = k_4$, and $a_2 = k_8$ are the velocity constants of the overall conversion of ethylene, acetylene, and ethane, respectively, the time t being replaced (in accordance with the equation of Vasil'ev, Kobozev, and Eremin) by the specific energy U/V [18].

On the basis of the experimental values of a_2 [19], a_3, and a_4 (for acetylene greatly diluted with hydrogen) [20], we selected numerical values of k_3, k_9, and k'_2. We then calculated from (1), (2), and (3) the curves of the decrease in C_2H_4 and the variation of the degrees of ethylene conversion to ethane and acetylene with changing specific energy. The calculations were made for three series of experiments — at current strengths of 225, 175, and 125 mA. The values of the constants are given in Table 2.

The dashed curves in Fig. 3 were plotted from these values of the constants. It will be seen that our scheme for ethylene conversion agrees with the experimental data.

We tried to analyze other schemes, particularly the scheme in which ethylene decomposes directly to the end products, characterized by the constant k_9. However, we could not obtain agreement between the curves and the experimental data for any value of k_3, k'_2, k_4, and k_8.

Study of the constants describing the experimental data reveals that the main duration of ethylene decomposition is acetylene formation. Furthermore, the overall conversion of ethylene takes place by a first-order reaction. These facts correspond most closely with a molecular pattern of acetylene formation, the activation of the initial molecules being due to electron impact:

$$C_2H_4 + e \rightarrow C_2H_4^* + e,$$
$$C_2H_4^* \rightarrow C_2H_2 + H_2.$$

The formation of ethane (a process with a lower velocity constant) may be represented by stagewise hydrogenation of ethylene by atomic hydrogen from the dissociation of hydrogen molecules or from some other source:

$$C_2H_4 + H \rightarrow C_2H_5,$$
$$C_2H_5 + H \rightarrow C_2H_6.$$

Owing to lack of experimental data, we cannot draw definite conclusions regarding the mechanism of methane formation. The CH_3 radicals which may be formed by the reaction

$$H + C_2H_5 \rightarrow 2CH_3$$

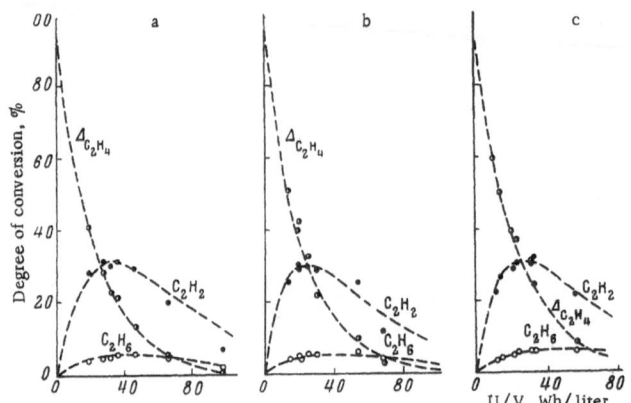

Fig. 3. Degree of overall conversion of ethylene ($\Delta_{C_2H_4}$) and degree of conversion of ethylene to the end products (γ), plotted versus the specific energy U/V. Current strength (mA): a) 225; b) 175; c) 125.

evidently recombine with atomic hydrogen; the molecule formed is stabilized on the wall because a triple collision is improbable under our conditions.

The presence of propane and propylene in the cracking products can be explained by the following reactions:

$$CH_3 + C_2H_5 \rightarrow C_3H_8,$$
$$CH_3 + C_2H_3 \rightarrow C_3H_6.$$

The C_2H_3 radical can be formed by direct dehydrogenation of an ethylene molecule, excited by an electron impact,

$$C_2H_4{}^* \rightarrow C_2H_3 + H,$$

or by the reaction

$$C_2H_4 + H \rightarrow C_2H_3 + H_2.$$

Furthermore, from the shape of the kinetic curves of $\gamma_{C_3H_8}$ and $\gamma_{C_3H_6}$ we can infer that the propylene is formed by dehydrogenation of propane

$$C_3H_8 \rightarrow C_3H_6 + H_2.$$

In a low-pressure glow-discharge zone the energy conditions may vary widely and the ethylene molecules may be activated not only by collisions with electrons with different energies, but also by thermal processes per se; therefore, these reactions may be accompanied by molecular polymerization of ethylene. We have already mentioned that at low discharge current strengths ethylene conversion is a second-order reaction. Milder conditions apparently assist predominance of ethylene polymerization in the overall conversion mechanism.

SUMMARY

1. Ethylene conversion in a low-pressure glow discharge may be described satisfactorily by the first-order kinetic equation of Vasil'ev, Kobozev, and Eremin. With decreasing power of the discharge we observe a change-over to a second-order reaction.

2. The main gaseous reaction products are acetylene and hydrogen. Methane, ethane, propane, and propylene are also formed.

3. A kinetic scheme has been postulated for ethylene conversion. According to this scheme, ethylene reacts in three ways: conversion to acetylene, hydrogenation to ethane, and conversion to other products, including methane, propane, propylene, and solid substances.

The equations corresponding to this scheme agree with the experimental data.

4. A mechanism is postulated for ethylene conversion.

LITERATURE CITED

1. Thenard, P., and A. Thenard, Compt. Rend., 76 : 1513 (1873).
2. Wilde, H., Ber., 7 : 357 (1874).
3. Collie, J., J. Chem. Soc., 87 : 1540 (1905).
4. Losanitsch and Jovitschitch, Ber., 30 : 138 (1897); 40 : 4664 (1907).
5. Dem'yanov, N. Ya., and N. D. Pryanishnikov, Zh. Russ. Fiz. Khim. Obshchestva, 58 : 462 (1926).
6. Lind, S. C., and G. Clockler, J. Am. Chem. Soc., 51 : 2811 (1929).
7. Andreev, D. N., Izv. Akad. Nauk SSSR, ser. khim., No. 5-6 : 1039 (1938).
8. Mund, W., and R. Goekelbergs, Ann. Soc. Sci., 1(65) : 149 (1951).
9. Mignonac, G., and R. V. St-Auny, Compt. Rend., 189 : 106 (1929).
10. Stratta, R., and E. Vernazza, Ind. Chimica, No. 6 : 133 (1931).
11. Szukiewicz, W., Roczniki Chem., 13 : 245 (1933).
12. Balandin, A. A., Ya. T. Éidus, and N. G. Zalogin, Dokl. Akad. Nauk SSSR, 4 : 132 (1934); Zh. Fiz. Khim., 6 : 389 (1935).
13. Éidus, Ya. T., and N. N. Nechaeva, Izv. Akad. Nauk SSSR, Otdel. Khim. Nauk, No. 1 : 153 (1940).
14. Volmar, J., and G. Hirtz, Bull. Soc. Chim., 49 : 684 (1931).
15. Reddy, M. P., and M. Burton, J. Am. Chem. Soc., 79 : 819 (1957).
16. Burton, M., and J. L. Magee, J. Chem. Phys., 23 : 2194 (1955).
17. Tsentsiper, A. B., E. N. Eremin, and N. I. Kobozev, Zh. Fiz. Khim., 37 : 835 (1963).
18. Vasil'ev, S. S., N. I. Kobozev, and E. N. Eremin, Zh. Fiz. Khim., 7 : 619 (1936).
19. Borisova, E. N., and E. N. Eremin, this collection, p. 33.
20. Borisova, E. N., and E. N. Eremin, this collection, p. 46.

KINETICS AND MECHANISM OF ETHANE CONVERSION
IN A GLOW DISCHARGE AT LOW PRESSURES

E. N. Borisova and E. N. Eremin

Under the effect of an electrical discharge, ethane may decompose in two ways; one of these involves splitting of the carbon–carbon bond and is a prerequisite for methane formation, while the other consists in detachment of hydrogen atoms from the ethane molecule to form ethylene and acetylene. Both reaction patterns can be effected by discharges of various form.

For ethane cracking we used a glow discharge at 0.7-1.7 mm Hg and a continuous-flow system. The apparatus and experimental procedure are described in [1]. We performed six series of experiments on ethane at the following discharge current strengths: 32, 75, 125, 175, 225, and 350 mA. The flow rate varied from 1 to 15 liters/h (0°C, 760 mm Hg).

Figure 1 gives the experimental results; the degrees of transformation of ethane to the end products $\gamma_{C_2H_2}$, $\gamma_{C_2H_4}$, $\gamma_{C_3H_8}$, and $\gamma_{C_3H_6}$ are plotted versus the specific energy U/V in Wh/liter.

It will be seen that acetylene is the principal component of the products condensed at the temperature of liquid nitrogen. The curve plotting $\gamma_{C_2H_2}$ versus the specific energy has a maximum, the height of which decreases with decreasing current strength in the discharge (18% at 350 mA, 14% at 175 mA, and 10% at 32 mA).

The reaction products also contain an appreciable amount of ethylene. The curves plotting the degree of conversion of ethane to ethylene also have maxima; with decreasing current strength the fall of the curves is increasingly retarded after the maximum and the height of the latter decreases from 6% at 350 mA to 4% at 32 mA.

It is noteworthy that the propane content in the reaction products is higher than in the case of ethylene decomposition. Whereas the maximum value of $\gamma_{C_3H_8}$ in ethylene cracking [2] was less than 3%, conversion of ethane to propane reaches 6.7% at 350 mA and decreases with decreasing current strength to 3% at 32 mA. The accumulation of propylene in the reaction products and the curves of the propylene content versus U/V and the current strength are similar to those for ethylene.

The reaction products and the unreacted hydrocarbon were condensed by our standard procedure in liquid-nitrogen traps after the discharge tube. It was therefore possible to follow the hydrogen and methane concentrations because they were not frozen at our experimental pressures. We performed several experiments in which the cracking gases were collected after the backing pump without precondensation; however, owing to losses in the pump, we did not determine the absolute amounts of the end hydrocarbons but their relative concentrations.

Thus, for a current strength of 225 mA in the discharge, we obtained the following composition of the cracking gases (Table 1).

46

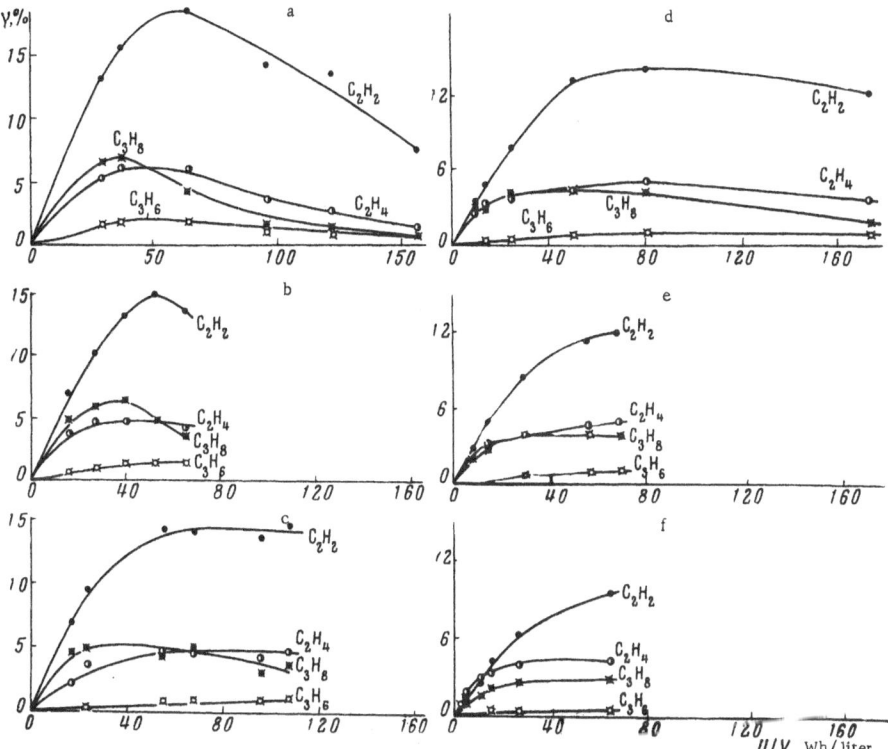

Fig. 1. Degree of conversion of ethane (γ) to the end products, plotted versus the specific energy U/V. Current strength (mA): a) 350; b) 225; c) 175; d) 125; e) 75; f) 32.

TABLE 1. Variation of End Gas Composition with Specific Energy

Amount of gas fed, liters/h	Electric parameters		Comp. of end gases, vol.%						
	U, W	U/V, Wh/liter	H_2	CH_4	C_2H_6	C_2H_4	C_2H_2	C_3H_6	C_3H_8
12.93	331	25.6	43.0	5.5	32.8	5.5	6.0	4.1	1.6
7.12	236	33.2	49.4	5.7	31.2	4.3	5.1	3.3	1.7
6.18	228	37.0	50.0	6.7	28.8	4.1	5.0	3.0	1.4
3.27	197	60.2	61.8	9.2	17.7	4.2	4.5	1.7	0.5

It is clear that the percentage content of methane in the products is fairly high (9 vol.%), whereas the conversion products of ethylene contained only slightly more than 3% methane at approximately the same specific energy. Furthermore, whereas the percentage contents of ethane, ethylene, acetylene, propane, and propylene fall with decreasing specific energy, the amounts of hydrogen and methane increase; this indicates that the latter are the gaseous end products of ethane conversion.

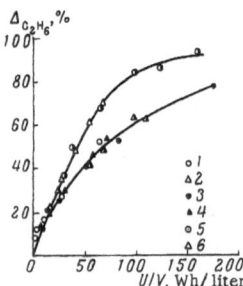

Fig. 2. Degree of conversion of ethane ($\Delta_{C_2H_6}$) versus specific energy U/V. Current strength (in mA): 1) 350; 2) 225; 3) 175; 4) 125; 5) 75; 6) 32.

In addition to the gaseous reaction products, solid substances were formed; these were deposited on the tube walls and the traps as solid tarry films, the color ranging from yellow almost to black as the discharge power or the specific energy increased. Ethane conversion to solid products was a few percent at low current strengths and 70-80% at high strengths, with overall conversion 80-90%. Figure 2 plots the overall ethane conversion $\Delta_{C_2H_6}$ (the ratio of the amount of reacted ethane to the total amount passed through the discharge) versus the specific energy in all the series of experiments. It will be seen that the data fall mainly into two groups. In the first group — at 225 and 350 mA — the degree of decomposition is greater than in the second group with smaller current strengths. Thus, although the degree of conversion of ethane can be represented as a function of U/V, this function is not unique in a wide range of current strengths. From the equation of Vasil'ev, Kobozev, and Eremin [3],

$$k = -\frac{1}{U/V} \ln (1 - \Delta_{C_2H_6}),$$

we calculated the velocity constants of the first-order reaction. It was found that in each series of experiments the mean values of the constants fall with decreasing current strength in the discharge (0.018 liter/Wh at 350 mA, 0.012 liter/Wh at 175 and 125 mA). At high current strengths, in each series of experiments the constants display little change, but at low current strengths they decrease with the specific energy, i.e., the reaction is not of the first order.

From an analysis of the kinetic curves for ethane decomposition, we postulated the following scheme:

$$C_2H_6 \xrightarrow{k_6} C_2H_2 \xrightarrow{k_4} \text{solid products, } H_2, C_3H_8, C_3H_6, CH_4,$$

in which ethane reacts in three parallel ways: formation of ethylene (k_2), direct transformation to acetylene, with no stable intermediate products (k_6), and formation of other products — CH_4, C_3H_8, C_3H_6, and solid substances (k_8). The ethylene formed from ethane reacts by the scheme in [2], but we omit hydrogenation to ethane (k_2') because the value of the constant is low and the amount of ethylene in the reaction products small, so that the ethane formed by hydrogenation is negligible in comparison with the initial amount.

Assuming that all the reactions are of the first order, we shall write the following differential equations for the rates of change of ethane concentration in the main products:

$$-\frac{d\,[C_2H_6]}{dt} = (k_2 + k_6 + k_8)\,[C_2H_6],$$

$$\frac{d\,[C_2H_4]}{dt} = k_2\,[C_2H_6] - (k_3 + k_9)\,[C_2H_4],$$

$$\frac{d\,[C_2H_2]}{dt} = k_6\,[C_2H_6] + k_3\,[C_2H_4] - k_4\,[C_2H_2].$$

Solution of this system gives the following equations for the degrees of conversion:

$$1 - \Delta_{C_2H_6} = e^{-a_1 U/V},$$

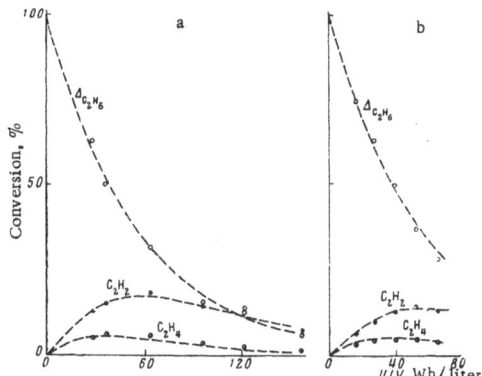

Fig. 3. Degree of overall conversion of ethane ($\Delta_{C_2H_6}$) and degrees of conversion of ethane to the end products, plotted versus the specific energy U/V. Current strength (in mA): a) 350; b) 225.

$$\gamma_{C_2H_4} = k_2 \left(\frac{1}{a_3 - a_2} e^{-a_2 U/V} + \frac{1}{a_2 - a_3} e^{-a_3 U/V} \right),$$

$$\gamma_{C_2H_2} = k_6 \left(\frac{b_0 - a_2}{(a_3 - a_2)(a_4 - a_2)} e^{-a_2 U/V} + \frac{b_0 - a_3}{(a_2 - a_3)(a_4 - a_3)} e^{-a_3 U/V} + \frac{b_0 - a_4}{(a_2 - a_4)(a_3 - a_4)} e^{-a_4 U/V} \right),$$

where $a_2 = k_1 + k_6 + k_8$, $a_3 = k_3 + k_9$, $a_4 = k_4$ are the constants of the overall conversion of ethane, ethylene, and acetylene, respectively, and

$$b_0 = a_3 + \frac{k_2 k_3}{k_6}.$$

From these equations we plotted the theoretical curves of ethane loss and conversion to various hydrocarbons versus the specific energy at high current strengths (350 and 225 mA), at which the reactions are of the first order. The dashed curves in Fig. 3 were plotted from the values of the constants in Table 2. Here a_2 and a_3 are the experimental values, a_4, k_3, and k_9 are calculated from kinetic schemes of ethylene decomposition [2], and k_2, k_6, and k_8 were selected so that their sum is equal to the experimental value of a_2.

It will be seen from Fig. 3 that the kinetic scheme agrees with the experimental value.

We made attempts to base the calculations on other kinetic systems, particularly a scheme in which all the acetylene is obtained sequentially from ethane only via ethylene. However, agreement with the experimental values was not obtained for any values of the constants. Nor were good results obtained for a scheme

TABLE 2. Variation of Velocity Constants with Current Strength

Current strength, mA	$a_2 = k_1 + k_6 + k_8$	$a_3 = k_3 + k_9$	$a_4 = k_4$	k_2	k_6	k_8	k_3	k_9
350	0.018	0.05	0.02	0.005	0.006	0.007	0.03	0.02
225	0.017	0.04	0.025	0.004	0.005	0.008	0.03	0.01

with no allowance for k_8. It can therefore be postulated that our scheme reflects the true picture of sequential-parallel conversion of hydrocarbons during ethane decomposition in a discharge.

From an analysis of this kinetic scheme, we can infer that the reaction has a radical mechanism. The presence of methane, propane, and propylene in the products excludes sequential molecular dehydrogenation of ethane to acetylene as the only or the predominant mechanism of ethane decomposition.

At low pressures a glow-discharge zone displays high electron temperatures and relatively low molecular temperatures. Therefore, the determinant process is not thermal activation of the molecules but excitation by electron impacts. The ethane molecules thus excited can decompose into radicals in two ways:

$$C_2H_6 \rightarrow 2CH_3, \qquad \Delta H^\circ = 86 \pm 2 \quad \text{kcal/mole,} \tag{1}$$

$$C_2H_6 \rightarrow C_2H_5 + H, \qquad \Delta H^\circ = 97 \pm 1 \quad \text{kcal/mole.} \tag{2}$$

The atomic hydrogen and CH_3 radical thus obtained react with the initial ethane molecules, as follows:

$$H + C_2H_6 \rightarrow C_2H_5 + H_2, \tag{3}$$

$$CH_3 + C_2H_6 \rightarrow C_2H_5 + CH_4, \tag{4}$$

to form ethyl radicals and the methane found in the reaction products. According to Steacie [4], the activation energies of (3) and (4) are less than 9 kcal/mole.

Ethylene may be formed by subsequent excitation of the ethyl radical in the discharge or by its decomposition, as follows:

$$C_2H_5 \rightarrow C_2H_4 + H. \tag{5}$$

The heat effect of this reaction is about 40 kcal/mole, and the activation energy is evidently less than 43 kcal/mole [4]. Ethylene may also be formed by molecular dehydrogenations of ethane:

$$C_2H_6 \rightarrow C_2H_4 + H_2. \tag{6}$$

To explain the direct formation of acetylene, we assume that the following reactions may occur:

$$C_2H_5 \rightarrow C_2H_3 + H_2, \tag{7}$$

$$C_2H_3 \rightarrow C_2H_2 + H, \tag{8}$$

the preceding radical activation being due to electron impacts.

Neglecting all the reactions in which two radicals participate (because their concentrations are low in comparison with the amount of initial ethane), and allowing for recombination at the walls, we get the following simplified mechanism:

$$C_2H_6^* \overset{k_1}{\rightarrow} 2CH_3,$$

$$C_2H_6^* \overset{k_2}{\rightarrow} C_2H_5 + H,$$

$$H + C_2H_6 \overset{k_3}{\rightarrow} C_2H_5 + H_2,$$

$$CH_3 + C_2H_6 \overset{k_4}{\rightarrow} C_2H_5 + CH_4,$$

$$C_2H_5^* \overset{k_5}{\rightarrow} C_2H_4 + H,$$

$$C_2H_5^{\bullet} \xrightarrow{k_7} C_2H_3 + H_2,$$

$$C_2H_3^{\bullet} \xrightarrow{k_8} C_2H_2 + H,$$

$$H \xrightarrow{k_{14}} \frac{1}{2} H_2,$$

in which the main role is played by activation of the molecules and radicals by electron impact, followed by decomposition of the excited particles. Processing of the results by the method of steady concentrations gives the following simple equation for the velocity of ethane decomposition:

$$- \frac{d[C_2H_6]}{dt} = 3(k_1 + k_2)[C_2H_6] + 2k_3 \frac{k_1 + k_2}{k_{14}} [C_2H_6]^2.$$

Thus, the reaction will be of the first or second order, depending on the conditions.

SUMMARY

1. We have studied the kinetics of ethane conversion in a glow discharge at low pressures.

2. It is shown that at high current strengths ethane conversion in the discharge obeys the first-order kinetic equation of Vasil'ev, Kobozev, and Eremin. With decreasing power of the discharge, the reaction is no longer of the first order.

3. The main gaseous product of ethane conversion is acetylene; smaller amounts of ethylene, methane, propane, and propylene are formed.

4. We have postulated a kinetic scheme of ethane conversion and derived corresponding equations which agree with the results. It is shown that the bulk of the acetylene is the product of direct decomposition of ethane without formation of ethylene as the essential intermediate product.

5. The direct formation of acetylene and the presence of methane, propane, and propylene in the reaction products indicate that ethane conversion in a glow-discharge has a radical mechanism.

Study of the mechanism showed that the reaction can be of the first or second order.

LITERATURE CITED

1. Borisova, E. N., and E. N. Eremin, this collection, p. 33.
2. Borisova, E. N., and E. N. Eremin, this collection, p. 40.
3. Vasil'ev, S. S., N. I. Kobozev, and E. N. Eremin, Zh. Fiz. Khim., 7:619 (1936); Khim. Tv. Topliva, No. 8:70 (1937).
4. Steacie, E. W. R., Atomic and Free Radical Reactions, 2nd ed. Reinhold, New York (1954).

KINETICS AND MECHANISM OF METHANE CONVERSION
IN A GLOW DISCHARGE AT LOW PRESSURES

E. N. Borisova and E. N. Eremin

The aim of the present paper is to elucidate the kinetics and mechanism of methane conversion in a glow discharge by a detailed study of electrocracking of methane and the behavior in an electric discharge of its main intermediate conversion products (ethane, ethylene, and acetylene).

The discharge apparatus and the reaction conditions [1] were selected so that we could accumulate considerable amounts of the less stable intermediate products (ethane and ethylene), as well as the most stable one (acetylene). Attempts to obtain considerable amounts of ethane by cracking methane in a glow discharge were made by Brewer and Kueck [2] and Tickner [3]; however, they found it necessary to employ deep cooling of the discharge tube by liquid nitrogen (and even solid nitrogen in some cases) to $-220°C$ and the yields were small.

We performed six series of experiments on methane at discharge-zone pressure 1.2-1.3 mm Hg at the following current strengths: 32, 75, 125, 175, 225, and 350 mA.

In each series the power of the discharge was approximately constant, while the flow rate of the gas, and, therefore, the specific energy, was varied.

The cracking products contained C_2H_6, C_2H_2, C_2H_4, C_3H_8, C_3H_6, H_2, and solid substances. Figure 1 plots γ, the conversion of methane to the end products, versus the specific energy for the six series of experiments. It will be seen that the main gaseous product of methane conversion is ethane. With increasing specific energy the $\gamma_{C_2H_6}$ curves pass through maxima which are displaced toward high U/V as the current strength falls; this is accompanied by an increase in the height of the maxima to 23%, i.e., nearly a quarter of the initial methane is converted to ethane.

The next most important product is acetylene. The curve plotting $\gamma_{C_2H_2}$ versus U/V also has a maximum, but, unlike ethane, the maxima increase with the current strength (i.e., with the discharge power) and are reached at high specific energies. The maximum value of $\gamma_{C_2H_2}$ is less than 16%.

The reaction products contain much less ethylene than ethane and acetylene. The curves plotting $\gamma_{C_2H_4}$ versus U/V have maxima, the positions of which coincide with those of $\gamma_{C_2H_6}$ at high current strengths, but are displaced toward the origin at low current strengths; the height of the maximum decreases from 4.5% at 350 mA to 1.7% at 32 mA.

The methane conversion products also contain appreciable amounts of propane and propylene. Whereas the $\gamma_{C_3H_6}$ curves are like the analogous curves for the ethane experiments, the values of $\gamma_{C_3H_8}$ are much smaller; for high current strengths the mean $\gamma_{C_3H_8}$ is about 1.5%, but increases with decreasing current strength and reaches 3.8% (at 75 mA).

A separate analysis for the acetylene derivatives showed that more than 13 hydrocarbons were present; of these, diacetylene, vinylacetylene, and methylacetylene were identified.

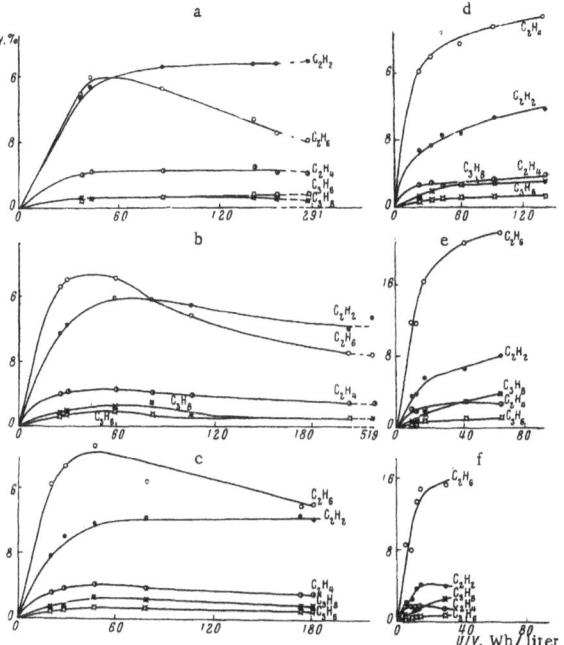

Fig. 1. Degree of conversion of methane to the end products (γ) versus specific energy U/V (P = 1.2-1.3 mm Hg). Current strength (mA): a) 350; b) 225; c) 175; d) 125; e) 75; f) 32.

In contrast to the other hydrocarbons, the unreacted methane did not condense in the traps at these low pressures, so that its content was determined indirectly. We determined the content of methane and hydrogen in the gas discharged from the backing pump (quantitative collection of the gas after the pump was impossible at the low flow rates in our experiments) and thus found the ratio $[H_2]/[CH_4]$. Knowing the qualitative and quantitative compositions of the reaction products condensed in the liquid-nitrogen traps, we can calculate the amount of hydrogen evolved during formation of these products from methane; we shall denote this by A. The amount of methane which should remain if solid reaction products were not formed will be denoted by B. Assuming that the solid products have approximately the composition $(CH_2)_n$, as found in analogous experiments by Yeddanapalli [4], we get the equation

$$\frac{[H_2]}{[CH_4]} = \frac{A + x}{B - x} ,$$

where x is the amount of methane decomposed to solid products.

Hence, the overall conversion is

$$\Delta_{CH_4} = \frac{V_{CH_4} - (B - x)}{V_{CH_4}} ,$$

where V_{CH_4} is the total amount of methane passed through the discharge.

Using this procedure, we calculated the constants for a series of experiments with current strength 225 mA by means of the equation of Vasil'ev, Kobozev, and Eremin [5]. The constants displayed little variation and the mean value was 0.026 liter/Wh, i.e., the decomposition constant of methane was greater than that of ethane under similar conditions (0.018 liter/Wh) [6].

At first sight, this fact is rather surprising because methane is considered as the most stable hydrocarbon; however, such observations have been reported before [7]. In our work, this result is confirmed by a comparison of the sums of the methane and ethane conversions to the gaseous reaction products: 35-42% for methane, but less than 30% for ethane [6].

It should be mentioned that we studied the effect of the traps' cold surfaces on the character and amount of the hydrocarbons formed during methane cracking. The experiments were performed without condensing the cracking gases at the temperature of liquid nitrogen (the end gas was collected after the pump and analyzed by chromatography); it was found that predominant formation of ethane does not depend on trap cooling, but is determined by the experimental conditions, particularly the low pressure and the design of the discharge tube.

Experiments in which the discharge-tube pressure was approximately twice as great as in the previous experiments were performed to determine the effect of pressure on the extent of conversion and the character of the products.

The experimental conditions and procedure, and the current strength in the discharge were the same as before. We shall not give the results because they are essentially the same as in Table 1. However, some discrepancies were observed. Thus, owing to the doubled pressure, the discharge power at the same current strengths increased markedly owing to the increased discharge potential. For example, at 350 mA the mean discharge power at 1.2-1.3 mm Hg was about 300 W, at 3 mm Hg it was 460 W, and in the experiments with low flow rates it was even more than 600 W. This difference in the power influenced the effectiveness of the discharge and thus explains the discrepancy in the results. For example, at 3 mm Hg the yield of C_2H_2 was high, while that of C_2H_6 was smaller. This difference is evidence of a specific trend; at even higher pressures (28-36 mm Hg) ethane is virtually absent in the reaction products [8].

From the kinetic curves for conversion of methane to various products, and bearing in mind the kinetic schemes for ethane and ethylene conversion, we established the following scheme of parallel and sequential reactions for methane conversion:

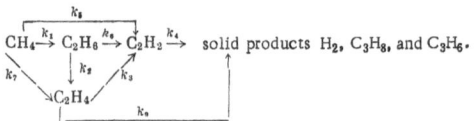

In this scheme, methane undergoes parallel decomposition in three ways: conversion to ethane (k_1) and direct transformation, i.e., without formation of stable intermediate products; to ethylene (k_7); and to acetylene (k_5). Conversion of the ethylene, ethane, and acetylene formed takes place by the kinetic schemes found by other investigators [1, 6, 9] for reactions of these hydrocarbons in a discharge. We disregarded the formation of methane from ethylene and ethane because their conversion to methane was low (as shown by the experimental data) and the amount of methane obtained was negligible in comparison with the initial amount.

Assuming that all the reactions are of the first order, we shall write the following differential equations for the rates of change of methane concentration in the main products:

$$- \frac{d[CH_4]}{dt} = (k_1 + k_5 + k_7)\,[CH_4],$$

$$\frac{d\,[C_2H_6]}{dt} = \frac{1}{2}\,k_1\,[CH_4] - (k_2 + k_6)\,[C_2H_6],$$

TABLE 1. Constants Adopted for the Calculations

Constant	$I = 225\,mA$ $P = 1-2$ mm Hg	$I = 350\,mA$ $P = 3$ mm Hg	Constant	$I = 225\,mA$ $P = 1-2$ mm Hg	$I = 350\,mA$ $P = 3$ mm Hg
$a_1 = k_1 + k_5 + k_7$	0.028	0.044	$a_3 = k_3 + k_9$	0.040	0.020
k_1	0.0185	0.021	k_3	0.030	0.020
k_5	0.006	0.016	k_9	0.010	0.000
k_7	0.0035	0.007	$a_4 = k_4$	0.020	0.004
$a_2 = k_2 + k_6$	0.018	0.009			
k_2	0.008	0.005			
k_6	0.010	0.004			

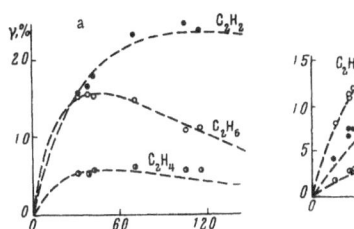

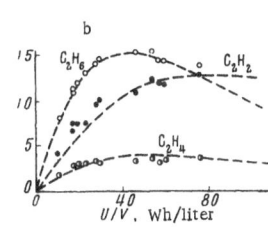

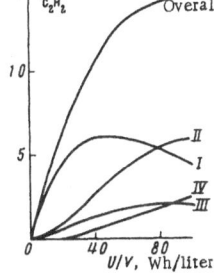

Fig. 2. Conversion of methane to end products (γ) versus the specific energy U/V. Curves plotted at current strength (mA): a) 350; b) 225.

Fig. 3. Comparison of different routes of acetylene formation from methane.

$$\frac{d\,[C_2H_4]}{dt} = \frac{1}{2}\,k_7\,[CH_4] + k_2\,[C_2H_6] - (k_3 + k_9)\,[C_2H_4],$$

$$\frac{d\,[C_2H_2]}{dt} = \frac{1}{2}\,k_5\,[CH_4] + k_6\,[C_2H_6] + k_3\,[C_2H_4] - k_4[C_2H_2].$$

Solution of this system gives the following equations for the degrees of conversion of the hydrocarbons:

$$1 - \Delta_{CH_4} = e^{-a_1 U/V},$$

$$\gamma_{C_2H_6} = \frac{1}{2}\,k_1\left(\frac{1}{a_2 - a_1}\,e^{-a_1 U/V} + \frac{1}{a_1 - a_2}\,e^{-a_2 U/V}\right),$$

$$\gamma_{C_2H_4} = \frac{1}{2}\,k_7\left(\frac{b_0 - a_1}{(a_2 - a_1)(a_3 - a_1)}\,e^{-a_1 U/V} + \frac{b_0 - a_2}{(a_1 - a_2)(a_3 - a_2)}\,e^{-a_2 U/V} + \frac{b_0 - a_3}{(a_1 - a_3)(a_2 - a_3)}\,e^{-a_3 U/V}\right),$$

$$\gamma_{C_2H_2} = \frac{1}{2}\,k_5\left[\frac{a_1^2 - a_1 b_1 + b_2}{(a_2 - a_1)(a_3 - a_1)(a_4 - a_1)}\,e^{-a_1 U/V} + \frac{a_2^2 - a_2 b_1 + b_2}{(a_1 - a_2)(a_3 - a_2)(a_4 - a_2)}\,e^{-a_2 U/V}\right.$$

$$\left. + \frac{a_3^2 - a_3 b_1 + b_2}{(a_1 - a_3)(a_2 - a_3)(a_4 - a_3)}\,e^{-a_3 U/V} + \frac{a_4^2 - a_4 b_1 + b_2}{(a_1 - a_4)(a_2 - a_4)(a_3 - a_4)}\,e^{-a_4 U/V}\right],$$

where a_1, a_2, a_3, and a_4 are the constants of overall conversion of CH_4, C_2H_6, C_2H_4, and C_2H_2, respectively, and

$$b_0 = a_2 + \frac{k_1 k_2}{k_7} \ , \quad b_1 = a_2 + a_3 + \frac{k_1 k_6 + k_3 k_7}{k_5} \ , \quad b_2 = a_2 a_3 + \frac{k_1 k_6 a_3 + k_3 k_7 b_0}{k_5} \ .$$

In these equations the reaction time is replaced, as before, by the specific energy U/V.

Repeating the procedure adopted in our previous work, we plotted the theoretical curves of the loss of methane and its degrees of conversion versus U/V. The values of the constants used for this purpose are given in the table.

In Fig. 2 the dashed lines are the theoretical curves. The agreement with the experimental data is very close.

We tried to obtain agreement between the experimental data and other schemes of methane conversion, particularly the sequential scheme of Cassel, disregarding k_5 or k_7 or k_6; however, agreement was not obtained for any values of the constants.

Thus the scheme of methane conversion describing the experimental kinetic curves indicates four possible paths of acetylene formation: I) directly from methane, with k_5; II) via ethane, bypassing ethylene, with k_1 and k_6; III) from ethylene obtained directly from methane (k_7 and k_3); IV) sequential formation of acetylene via ethane and ethylene as intermediate products (k_1, k_2, k_3). When $\gamma_{C_2H_2}$ was calculated from the equations corresponding to each of these paths (Fig. 3), we found that even under conditions favoring ethane formation the bulk of acetylene formed in methane cracking originates via the first reaction route; the second and third routes are of less importance, and only about one tenth of the acetylene is formed sequentially via ethane and ethylene, i.e., by Cassel's scheme. This will be seen from Fig. 3, which also shows the overall conversion of methane to acetylene.

A change in the discharge conditions may be accompanied by a change in the ratios between the constants. It is highly probable that under more severe conditions and high pressures and current strengths the main role will be played by radical processes of direct conversion of methane to acetylene.

SUMMARY

1. We have studied the kinetics of methane conversion in a glow discharge at low pressures.

2. It is shown that the reaction obeys the first-order equation of Vasil'ev, Kobozev, and Eremin.

3. It has been established that under our experimental conditions the rate of methane conversion is greater than that of ethane conversion.

4. It is shown that ethane can be formed as the main product of glow-discharge cracking of methane without deep cooling of the discharge tube.

5. We have postulated a kinetic scheme of methane conversion, with three principal routes. Satisfactory agreement with the experimental data was obtained for the results calculated from the equations corresponding to our scheme, using the velocity constants obtained in our previous work. It is shown that the bulk of the acetylene obtained is formed by direct conversion of methane, without the formation of stable intermediate products.

6. It is shown that the mechanism of the sequential conversion (methane → ethane → ethylene → acetylene) is untenable for the reaction of methane in a glow discharge.

LITERATURE CITED

1.   Borisova, E. N., and E. N. Eremin, this collection, p. 33.
2.   Brewer, A. W., and P. D. Kueck, J. Phys. Chem., 35:1231 (1931).
3.   Tickner, A. W., Can. J. Chem., 39:87 (1961).
4.   Yeddanapalli, L. M., J. Chem. Phys., 10:249 (1942).
5.   Vasil'ev, S. S., N. I. Kobozev, and E. N. Eremin, Zh. Fiz. Khim., 7:619 (1936).

6. Borisova, E. N., and E. N. Eremin, this collection, p. 46.
7. Tsentsiper, A. B., E. N. Eremin, and N. I. Kobozev, Zh. Fiz. Khim., 37 : 1063 (1963); Dokl. Akad. Nauk SSSR, 141 : 378 (1961).
8. Borisova, E. N., and E. N. Eremin, Zh. Fiz. Khim., 36 : 1261, 2334 (1962).
9. Borisova, E. N., and E. N. Eremin, this collection, p. 40.

THE EFFECT OF THE ELECTRODE MATERIAL
AND THE SOOT CONTENT OF THE FEEDSTOCK ON ACETYLENE
MANUFACTURE BY DECOMPOSITION OF A LIQUID ORGANIC
FEEDSTOCK IN HIGH-VOLTAGE DISCHARGES

N. S. Pechuro, E. Ya. Grodzinskii, and O. Yu. Pesin

The decomposition of liquid organic products in high-voltage discharges (arc or pulse discharges) is accompanied by formation of solid decomposition products (soot) in addition to cracking gas. Both the absolute and relative yields of these products depend largely on the character of the feedstock [1, 2]. It has been established that stoppages in electrocracking are mainly due to the formation of soot bridges between the electrodes. If the feedstock circulation between the interelectrode gap is sufficiently great (5-6 m/sec), the solid decomposition products are removed from the reactor, thus ensuring continuity of the process [3]. This procedure also provides the necessary prerequisites for increasing the soot content in the feedstock without the danger of short circuits or coking-up of the apparatus. It was therefore of interest to find whether an increase in the soot content in the feedstock influences the main indices of the process, particularly the composition and yield of the gas and the specific power consumption. The apparatus and method used for this work were the same as in [2].

The main difficulty was to determine the concentration of solid particles in the discharge zone; under our conditions direct determination (by taking samples and separating and weighing the soot) was made difficult by the small volume of the feedstock circulating in the apparatus. Furthermore, the accuracy of such a determination is low, particularly in the initial period when the soot content in the feedstock is relatively small.

To determine the concentration of solid decomposition products in the feedstock, we used a sufficiently accurate method in which the calculations are based on laws already derived [1] for the relation between gas composition and the yield of decomposition products.

Starting from these laws we find the amount of soot formed per gram of cracking gas, using the formula

$$\frac{C_{\%}}{V_{\%}} = \frac{A(A + \gamma_g)}{(A + \gamma_g)\gamma_g} = \frac{A}{\gamma_g} , \frac{\text{grams of soot}}{\text{gram of gas}} . \tag{1}$$

In (1),

$$A = K\, 0.09\, ([H_2] + [C_2H_2] + 2\,[C_2H_4] + 2\,[CH_4]) - 0.535\,([CH_4] + 2\,[C_2H_2] + 2\,[C_2H_4]),$$

where the content of H_2, C_2H_2, C_2H_4, and CH_4 in the cracking gas is expressed in vol.%; K is the weight ratio of carbon to hydrogen in the cracking gas; $C_{\%}$ and $V_{\%}$ are the yields of soot and gas, respectively, per unit of decomposed feedstock, and γ_g is the weight of 1 liter of gas.

The amount of soot formed per liter of gas will be

$$\frac{A\gamma_g}{\gamma_g} = A, \frac{\text{grams of soot}}{\text{liter of gas}}.$$

Knowing the amount of gas formed by a given time (V_g^t) and the volume of feedstock circulating (V_c), we can calculate its soot content (B) from the formula

$$B = \frac{AV_g^t}{V_c}, \frac{\text{grams of soot}}{\text{liter of feedstock}}. \tag{2}$$

Since the volume of feedstock circulating in the system may gradually decrease (owing to decomposition), we must make a corresponding correction to V_c. The feedstock consumption per gram of cracking gas can be expressed as $1/V_\%$ (g/g), or as

$$\frac{\gamma_g}{\gamma_c V_\%}, \frac{\text{cm}^3 \text{ of feedstock}}{\text{liter of gas}},$$

where γ_c is the weight of the feedstock in g/cm^3.

The volume of feedstock decomposed by a time t can then be found from the formula

$$\Delta V_c = \frac{\gamma_g V_g^t}{V_\% \gamma_c}, \text{cm}^3. \tag{3}$$

Hence, the amount of feedstock circulating in the system will be

$$V_c = V_c^H - \Delta V_c = V_c^H - \frac{\gamma_g V_g^t}{\gamma_c V_\%}, \text{cm}^3, \tag{4}$$

where V_c^H is the volume of liquid product added to the system.

Putting the expression for V_c into (2) and making transformations, we finally get

$$B = \frac{AV_g^t}{V_c^H - \dfrac{\gamma_g V_g^t}{\gamma_c V_\%}} = \frac{AV_g^t}{V_c^H - \dfrac{\gamma_g V_g^t (A + \gamma_g)}{\gamma_c \gamma_g}} = \frac{AV_g^t \gamma_c}{V_c^H \gamma_c - V_g^t (A + \gamma_g)}. \tag{5}$$

The effect of the soot concentration on the process was studied on a hydrocarbon compound (benzene) and a petroleum product with initial boiling point 150°C and final boiling point 255°C. The experimental conditions were the same in both cases*: l_{ig} = 1.0 mm; U_{nl} = 15.0 kV; I_w = 57 mA; t_c = 25°C, and ω = 4.8 liters per min.

With benzene as the feedstock, the values of A and γ_g were calculated for a gas containing 43.6% C_2H_2, 1.2% $C H_{2n}$, 0.8% C_nH_{2n+2}, and 54.4% H_2. A was found to be 0.56 and γ_g 0.599 g/liter.

With V_c^H (for benzene) 0.8 liter, and γ_c 879 g/liter, the soot concentration can be calculated as follows:

$$B = \frac{0.56 \cdot 879 \, V_g^t}{0.8 \cdot 879 - V_g^t (0.56 + 0.599)} = \frac{492 \, V_g^t}{703 - 1.16 \, V_g^t}, \text{g/liter}. \tag{6}$$

Table 1 gives the values thus obtained for the soot concentration in benzene, and the main electrocracking indices corresponding to these values. It will be seen from Table 1 that a change in the soot content in the feedstock had no effect on the cracking-gas composition.

---

*l_{ig} is the interelectrode gap; U_{nl} is the no-load potential of the power supply; t_c is the feedstock temperature; ω is the amount of feedstock circulating; and I_w is the working current.

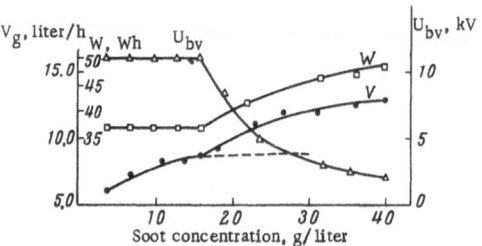

Fig. 1. Effect of the soot concentration in benzene on the
electrocracking indices.

TABLE 1. Benzene Decomposition at Various Soot Concentrations in the Discharge Gap
(U_p = 1000 V, N = 57 VA)

Time since begin. of expt., min	V_g^t, liters	B, g/liter, calc. data	V, liters/h	Gas comp., vol.%				U_{bv}, kV	d, %	q_n, $\frac{kWh/m^3}{C_2H_2}$ (•)
				C_2H_2	C_nH_{2n}	H_2	C_nH_{2n+2}			
25	5.45	3.87	6.3	33.9	0.9	64.2	1.0	11	65	—
50	10.77	7.72	7.4	37.4	1.0	60.5	1.1	11	65	—
99	21.26	15.51	8.3	42.6	1.1	55.4	0.9	11	65	11.3
146	31.54	23.40	11.0	43.4	1.0	54.6	1.0	5	74	8.8
194	41.62	31.40	12.1	43.6	1.2	54.4	0.8	3	82	8.7
240	51.62	39.60	12.9	43.3	1.0	54.5	1.2	2	85	8.6

Note: U_p is the potential drop in the discharge gap; U_{bv} is the breakdown voltage of
the gap; N is the power of the arc discharge; d is the relative duration of arc combustion (as % of the half-cycle); and q_n is the power consumption per m^3 of C_2H_2 (•) NTP.

The decrease in the acetylene content in the gas (at the beginning of the experiment) at low soot concentrations (up to about 15 g/liter) is due to its absorption by the liquid phase; such absorption ceases after the evolution of about 25-30 liters of cracking gas. Despite the subsequent continuing increase in the feedstock's soot content (up to almost 40.0 g/liter), the gas composition remains virtually constant.

It will be seen from Table 1 and the figure that the feedstock's soot content has a marked effect on the
yield of cracking gas. The curve plotting the gas yield versus the feedstock's soot content has an inflection at a
point corresponding to B ≈ 15.5 g/liter. The increase in the gas yield, which characterizes the left-hand branch
of the curve, is due to gradual saturation of the liquid phase by gaseous decomposition products. Graphical ex-
trapolation of this sector (cf. the dashed line in the figure) shows that with V_g^t ≈ 30 liters (i.e., with B ≈ 20 g per
liter) the gas yield should be stabilized. The arc parameters, particularly the breakdown voltage and the energy
evolved in the discharge gap (W), remain virtually constant. When the soot concentration reaches 15.5 g/liter,
the gas stability decreases markedly, the breakdown voltage falls, and, therefore, the arc combustion time increases. This leads to an increase in the energy evolved in the discharge channel and a proportional increase in
the gas yield.

It will be seen from Table 1 that the soot concentration in the feedstock has no effect on the working potential of the arc discharge. Therefore, the difference between the working voltage and breakdown voltage may
be greatly reduced by the gradual accumulation of solid decomposition products in the feedstock. At a soot concentration in benzene of, say, ~3.9 g/liter, the value of ΔU ($\Delta U = U_{bv} - U_p$) is ~10 kV, but with B = 39.6

g/liter only ~1 kV. Thus we can use for electrocracking plants transformers with relatively low no-load potentials — a fact of great importance for the planning of industrial units.

If the transformer retains its U_{nl} value at the given soot concentration, we can greatly increase the interelectrode gap. In particular, with B = 39.6 g/liter and U_{nl} = 15 kV, the interelectrode gap was increased from 1.0 to 3.5 mm.

Below we give the main indices of benzene decomposition with l_{ig} = 3.5 mm:

I_p = 50 mA; $\qquad$ N = 99.0 VA;

U_p = 1980 V; $\qquad$ d = 60%;

V_g = 17.0 liters/h; $\qquad$ q_n = 8.0 kWh/m^3(NTP) C_2H_2.

C_2H_2 = 43.5 vol.%;

These data confirm our previous inference [3] to the effect that an increase in the interelectrode gap (in the given case, owing to the increased soot concentration in the feedstock) not only raises the productivity of the apparatus, but also reduces the power consumption per m^3(NTP) C_2H_2.

We studied the effect of the soot concentration in the feedstock on the process indices for the case of decomposition of a petroleum product with initial boiling point 150°C and final boiling point 255°C.

Decomposition of this feedstock is accompanied by formation of about 44 g of soot per m^3(NTP) of gas, i.e., only about one fifth as much as for benzene. Table 2 gives the results obtained for decomposition of the petroleum product with increasing soot content in the feedstock. To increase the yield of gas and thus increase the deposition of soot, the interelectrode gap was increased to 2.0 mm.

It will be seen from Table 2 that after the evolution of about 40 liters of gas, the composition remains virtually unchanged (as in the case of benzene). At the same time the yield of gas is stabilized. This is due to the low soot concentration in the feedstock (only about 4.5 g/liter, even after the evolution of 80 liters of cracking gas). As in the case of benzene, such a very small content of decomposition products has no effect on the discharge parameters and, therefore, no influence on the productivity of the apparatus.

With the low capacity of our laboratory apparatus, an increase in the petroleum product's soot concentration to about 15 g/liter would have required protracted experiments. From the technological aspect, interest attaches to a way of obtaining a given soot concentration by adding it from outside.

To verify the feasibility of such addition, we added to the petroleum product soot obtained by its decomposition in an arc discharge. The soot was kept in a desiccator and contained about 71.5% of occluded liquid feedstock. A suspension of soot in the petroleum product (containing 15.7 g/liter) was obtained by mixing and then added to the system while the pump was working. The experiments showed that even at the outset the yield of gas was much higher than for the petroleum product itself. After saturation of the gas with soot, its composition and yield are stabilized (because the subsequent increase in soot concentration due to decomposition of the feedstock is very small). At this time the process has the following indices:

B = 15.7 + 2.4 = 18.1 g/liter,*

V = 17.2 liters/h; $\qquad$ U_p = 1060 V;

$[C_2H_2]$ = 37.0 vol.% $\qquad$ U_{bv} = 1060 V;

$[C_nH_{2n}]$ = 7.1 vol.% $\qquad$ d = 85%;

$[H_2]$ = 53.1 vol.% $\qquad$ N = 60.4 VA;

$[C_nH_{2n+2}]$ = 2.8 vol.% $\qquad$ q_n = 8.1 kWh/m^3(NTP) C_2H_2.

---

*2.4 g is the amount of soot formed by decomposition of the feedstock.

TABLE 2. Decomposition of a Petroleum Product at Various Soot Concentrations in the Discharge Gap (U_p = 1060 V, U_{bv} = 12 kV, N = 60.4 VA, d = 77%)

V_g^t	B, g/liter, calc.data	V, liters/h	Gas composition, vol.%				kWh/m³* $\frac{q_n}{C_2H_2}$
			C_2H_2	C_nH_{2n}	H_2	C_nH_{2n+2}	
5.2	0.3	7.2	30.2	6.2	60.4	3.2	--
10.4	0.6	10.8	32.5	6.5	58.0	3.0	--
20.6	1.1	13.5	35.2	6.7	55.2	2.9	--
30.9	1.7	15.0	36.4	7.0	53.8	2.9	--
41.0	2.3	15.2	36.9	7.4	53.0	2.7	8.3
51.0	2.9	15.5	37.0	7.2	53.0	2.8	8.2
61.0	3.4	15.3	36.8	7.4	53.2	2.6	8.3
71.0	3.9	15.2	37.1	7.4	52.8	2.7	8.3
81.0	4.5	15.4	37.0	7.2	53.0	2.8	8.2

*NTP.

TABLE 3. Gas Composition and Yield for Electrodes of Various Materials

Material	mp, °C	bp, °C	Gas yield, liters/h	Gas comp., vol. %				Total unsaturated hydrocarbons, vol.%
				C_2H_2	C_nH_{2n}	H_2	C_nH_{2n+2}	
Cd	321	767	11.9	35.6	7.9	53.6	2.9	43.5
Zn	419	907	11.7	35.4	7.8	53.6	3.2	43.2
Bi	271	1560	12.1	35.7	8.0	53.5	2.8	43.7
Dural	600	2000	15.0	35.2	7.7	53.8	3.3	42.9
Brass.	1000	2500	15.6	36.1	7.1	53.8	3.0	43.2
Steel 3	1700	3200	15.9	37.0	7.2	53.1	2.7	44.2
Graphite	—	3500 (subl.)	17.1	36.4	7.0	53.5	3.1	43.4

TABLE 4. Effect of the Electrode Material on the Power Consumption in Cracking and on Specific Erosion

Material	U_p, V	N, VA	d, %	q_n, kWh/m³* C_2H_2	Electrode erosion	
					g/m³* gas	g/m³* C_2H_2
Cd	1200	88.3	77	12.4	6.95	19.5
Zn	1250	71.3	77	13.2	5.60	15.8
Bi	1200	68.3	77	12.2	21.20	59.5
Dural	1250	71.2	77	10.3	0.51	1.26
Brass.	1250	71.2	77	9.7	0.94	2.60
Steel 3. . . .	1150	65.6	77	8.6	0.40	1.08
Graphite . . .	1200	68.3	84	9.2	7.83	21.50

*NTP.

A comparison of our results with the data in Tables 1 and 2 shows that the laws governing benzene decomposition hold true for the petroleum product.

Addition of lampblack to the petroleum product did not have a beneficial effect; the soot was rapidly deposited, blocking the pump and the pipes.

It must be emphasized that under our conditions an increase in soot concentration to more than 25-30 g/liter was undesirable because it increased the feedstock's viscosity, making transport difficult in the pipes.

The studies were performed with the same electrode material, Steel 3. Fester et al. [4] had established that in the decomposition of liquid organic products in a low-voltage arc the electrode material has an effect on the yield of unsaturated hydrocarbons, including acetylene. Equivalent data for high-voltage arcs have not been published. We therefore studied how the electrode material influences the process indices and also determined the erosion of various materials in a high-voltage arc discharge in a liquid dielectric.

The electrodes were made from materials with various boiling and melting points — from cadmium to graphite. The experiments were performed on a petroleum product (initial boiling point 150°C, final boiling point 255°C) under the following conditions:

$$l_{ig} = 1.0 \text{ mm}; \qquad U_{nl} = 15 \text{ kV};$$
$$\omega = 4.8 \text{ liters/h}; \qquad t_c = 25°C.$$
$$I_p = 57 \text{ mA};$$

It will be seen from Table 3 that the electrode material has practically no effect on the gas composition. The observed reduction in the acetylene concentration in the case of low-melting materials (Cd, Zn, Bi) is very small and within the experimental error.

Regarding their effect on the gas yield, all the electrode materials can be divided into two groups: a group with relatively low melting and boiling points (Cd, Zn, Bi) and a group with high mp and bp (dural, brass, steel, and graphite). For the first group the mean acetylene yield is 4.0 liters/h, and for the second group about 5.5 liters/h, i.e., 37% higher.

Analysis of the oscillograms showed that the same pattern is observed for the power consumption per $m^3(NTP)$ C_2H_2 (Table 4); for the first group it is 12-13 kWh/$m^3(NTP)$ C_2H_2, for the second group 9-10 kWh/m^3 (NTP) C_2H_2.

Minimum power consumption was observed in the case of steel and graphite electrodes. The use of graphite electrodes in place of metal electrodes markedly stabilizes the arc and increases its burning time (from $7.7 \cdot 10^{-2}$ to $8.4 \cdot 10^{-2}$ sec) and, therefore, the gas yield per unit time.

The electrode material has the most marked effect on erosion in the arc discharge. The erosion of low-melting metals is very high: for bismuth (mp 271°C) it reaches 21.2 g/m^3(NTP) of gas. Materials with melting points in the range 600-1700°C undergo relatively little erosion [0.5-1 g/m^3(NTP) of gas]. The unusually high erosion of graphite is evidently due not to its physical properties, but to the low mechanical strength of our material.

For industrial plants, it is desirable to use graphite electrodes because they stabilize the arc, ensure a relatively low power consumption per m^3(NTP) C_2H_2, and increase the gas yield per unit time.

SUMMARY

1. A soot content of about 15 g/liter in the feedstock is accompanied by a marked reduction in the breakdown voltage; with the use of an ac arc, this leads to an increase in the energy evolved in the discharge gap and an increase in the gas yield.

2. The composition of the cracking gas, the working potential, and the power consumption per m^3(NTP) C_2H_2 are independent of the soot concentration in the feedstock.

3. By keeping the soot concentration in the discharge gap at about 30 g/liter, we can reduce severalfold the power supply voltage at the given interelectrode gap, or make a corresponding increase in the latter while keeping the transformer's output voltage unchanged.

4. The effect of soot on the process indices is independent of the character of the feedstock cracked.

5. The necessary soot concentration in the feedstock can be obtained by adding solid products obtained by its decomposition and containing a certain amount of the liquid phase.

6. The electrode material has little effect on the composition of the cracking gas obtained by decomposition of a liquid organic feedstock in a high-voltage arc discharge.

7. The use of electrodes with high melting points (above 1000°C) and high boiling points (above 2000°C) leads to an increase in the overall yield of unsaturated compounds per unit time, a reduction in the power consumption per m^3(NTP) C_2H_2, and a reduction in the erosion of the material.

It must be emphasized that these conclusions apply to the arc discharge power used in the present work and not necessarily to other power values.

LITERATURE CITED

1. Pechuro, N. S., E. Ya. Grodzinskii, and O. Yu. Pesin, Symposium: Aspects of the Electrical Processing of Materials, No. 4, Izd. Akad. Nauk SSSR, Moscow (1962), p. 192.
2. Pechuro, N. S., E. Ya. Grodzinskii, and O. Yu. Pesin, Gaz. Prom., No. 2: 47 (1963).
3. Grodzinskii, E. Ya., N. S. Pechuro, and O. Yu. Pesin, Symposium: Arc Processing of Metals., Izd. Akad. Nauk SSSR, Moscow (1963), p. 100.
4. Fester, G. A., E. A. Martinuzzi, and A. Ricardi, Erdöl u. Kohle, 10: 840 (1957).

CODECOMPOSITION OF LIGHT AND HEAVY HYDROCARBONS
IN HIGH-VOLTAGE DISCHARGES

N. S. Pechuro and E. K. Starostin

A characteristic feature of liquid hydrocarbon decomposition in electric discharges is that considerable amounts of energy can be evolved in a small reaction space, with simultaneous quenching of the reaction products in the liquid medium. Under these conditions we can obtain gas with a high acetylene concentration (25-32 vol.%). When such processes are effected in unsteady electric discharges (as in the Tatarinov method [1]) the main electrodes and the intermediate current-conducting contacts undergo fairly considerable erosion, thus making it very difficult to obtain an appropriate layout of the apparatus. The use of a high-voltage arc [2,3] is accompanied by other difficulties because, owing to the high dielectric properties of the petroleum products cracked, it is difficult to employ discharge gaps permitting normal functioning of the process.

We have proposed an essentially new technological alternative for the process [4], permitting an increase in the interelectrode gap while retaining the previous feature of rapid quenching of the decomposition products. This method provides for preferential decomposition of pre-evaporated light petroleum products, fed to the discharge zone of a high-voltage arc, surrounded by a high-boiling feedstock. This enables us to increase the discharge gap, because the arc is struck by breakdown of the vapor phase, and to quench the decomposition products in the high-boiling petroleum products in the liquid phase.

The layout of the laboratory apparatus is shown in Fig. 1.

The main component of the apparatus for codecomposition of light and heavy hydrocarbons in a high-voltage arc was a molybdenum glass reactor 1. Electrodes 3 and 4, passing through high-voltage insulators 5, were located in the lower and upper parts of the apparatus. The lower, hollow electrode 3 is stationary; the upper electrode can be moved by an adjuster 6. A potential ($U_{nl} \simeq 10$ kV) was fed to the electrodes from a TG-1020 high-voltage transformer 7.

The low-boiling feedstock was circulated in the reactor. The heavy petroleum product was placed in the reactor and heated to about 140-160°C before the experiment was begun; this excluded condensation of the light feedstock, fed to the discharge zone via the lower, hollow electrode. The heavy petroleum product was heated to the required temperature by an electric heater 8. The bulk of the soot formed remained in the heavy product and was filtered off after the experiment.

The pre-evaporated low-boiling product with final boiling point 120-130°C decomposed in the high-voltage arc, and the vapor—gas mixture thus obtained was quenched in the heavy organic feedstock. The gas was fed through the condenser 9, in which the undecomposed low-boiling petroleum product condensed, and then through the scrubber 10 to remove entrained soot.

Very marked separation of the light fraction from the gas current took place in dry-ice trap 13. The amount of gas formed was measured by a dry-gas counter 14. An overpressure (about 100 mm Hg) was maintained in the system. The condensed low-boiling product was collected in a receiver 15 and then fed by a

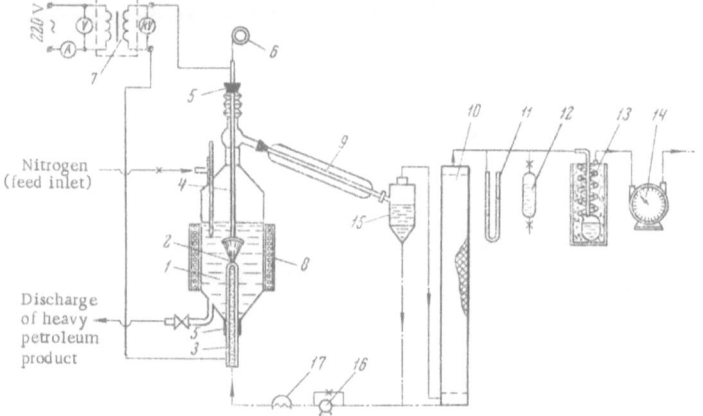

Fig. 1. Layout of the laboratory apparatus for codecomposition of heavy and light hydrocarbons. 1) Reactor; 2) arc zone; 3) lower electrode; 4) upper electrode; 5) high-voltage
insulators; 6) electrode-shifting device; 7) high-voltage transformer; 8) electric heater;
9) condenser; 10) scrubber; 11) water gauge; 12) gas pipet; 13) trap; 14) gas meter; 15)
vessel for low-boiling product; 16) circulation pump; 17) heater—evaporator.

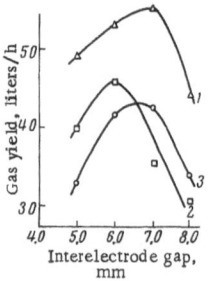

Fig. 2. Gas yield plotted versus amount of
low-boiling feedstock and interelectrode
gap. Amount of feedstock (in liters/h):
1) 2.0; 2) 1.2; 3) 0.6.

circulation pump 16 to a heater—evaporator 17. The feedstock,
heated to 20-30°C above the boiling point in the evaporator,
was then fed to the reaction zone of the high-voltage arc via
the lower, hollow electrode.

The electric circuit included a power supply 7 and the
necessary measuring instruments.

Table 1 gives the characteristics of the feedstocks. In addition to these, we used the following individual hydrocarbons:
n-hexane (low-boiling feedstock), n-decane, and n-undecane
(high-boiling feedstock).

In the first stage we studied the effect of the interelectrode gap and of the amount of low-boiling product fed to the
reaction zone on the gas yield and composition. We also determined the power indices under optimal conditions and compiled
the corresponding mass balances.

It will be seen from Table 2 that the character of the
feedstock has little effect on the gas yield and composition.
This may be due to the fact that the material cracked consisted of individual paraffinic hydrocarbons and low-
boiling petroleum products with considerable contents of such compounds (this is indicated by the low iodine
numbers, the aniline point, and the total amount of sulfonatable substances).

It has been established that the productivity of the apparatus (gas yield in liters/h) changes with the interelectrode gap and the amount of low-boiling product fed (Fig. 2). For example, by increasing the feedstock
from 0.6 to 2.0 liters/h at an interelectrode gap of 6-7 mm, the gas yield can be increased from 42 to 54
liters/h. Under our experimental conditions, with a further increase in the feed of low-boiling products per

TABLE 1. Characteristics of the Initial Petroleum Products Used for Codecomposition in a High-Voltage Arc

Characteristic	Low-boiling feedstock			High-boiling feedstock		
	1st gasoline fraction	2nd gasoline fraction	Petroleum ether	Diesel oil	Vacuum gas-oil	Transformer oil
Density d_4^{20}, g/cm^3 . . .	0.7064	0.7383	0.6760	0.8481	0.9120	0.9030
Refractive index n_D^{20}. . .	1.4043	1.4193	1.3817	1.4762	1.5088	1.4964
Aniline point, °C.	66.5	67.0	–	–	–	–
Iodine number (Kaufmann's method) .	7.0	7.7	2.0	21.8	28.3	22.6
Total sulfonatable compounds (Kattwinkel method), vol.%	12.0	20.0	1.2	30.5	–	33.3
Flash point, °C	–	–	–	95	185	152
Ignition point, °C	–	–	–	106	200	173.5
Engler distillation, °C						
ibp.	72	98	72	212	180/10 mm Hg	81/4 mm Hg
distillate 10%	76	101	76	237		
20%	79	103	77	246		
30%	82	104	79	255		
40%	85	105.5	80	265		
50%	88	107.0	81	277		
60%	91	109	83	289		
70%	94	112	85	303		
80%	100	115	87	321		
90%	105	121	90	338		
fbp.	127	139	94	340		

Note: The following contents were established by gas–liquid chromatography in the petroleum ether: 0.52% C_5, 18.6% C_6, 63.4% C_7, 17.48% C_8.

TABLE 2. Composition of the Gas Obtained by Codecomposition of Some Organic Products

Initial feedstock		Gas composition, vol.%							
Light	Heavy	H_2	CH_4	C_2H_6	C_2H_4	C_3H_8	C_3H_6	C_2H_2	C_4H_8
Petroleum product I (ibp 72°C, fbp 127°C)	Diesel oil	50.5	6.9	1.0	10.7	0.3	3.3	26.4	0.9
Petroleum product II (ibp 98°C, fbp 139°C)	Diesel oil	50.4	7.1	1.3	12.3	0.9	3.7	25.4	0.9
Petroleum ether	Transformer oil	52.1	9.3	1.0	6.9	–	3.8	26.2	0.7
n-Hexane	Vacuum gas-oil	52.5	6.2	0.5	12.9	0.2	1.1	25.8	0.8
n-Hexane	n-Undecane	53.0	6.2	0.8	14.5	0.3	1.3	23.4	0.5
n-Hexane	n-Decane	51.0	6.3	0.9	14.3	0.2	1.7	24.9	0.7

Note: $U_{nl} \simeq 10$ kV, N $\simeq 200$ W, $I_p = 0.026$ A, $l_{ig} \sim 5$ mm; amount of low-boiling feedstock fed ~1.0 liter/h. The gases were analyzed by gas adsorption chromatography.

unit time (to more than 2.0 liters/h) we observed disruption of the arc, due to the inadequate power supply. In these experiments the low-boiling feedstock was n-decane and the high-boiling feedstock was n-undecane.

A change in the interelectrode gap and the amount of low-boiling feedstock has little effect on the gas composition (cf. Table 3).

TABLE 3. Change in Gas Composition with the Interlectrode Gap and the Amount of Low-Boiling Feedstock

Amount of low-boiling feedstock, liters/h	Interelec. gap, mm	Gas composition, vol.%							
		H_2	CH_4	C_2H_6	C_2H_4	C_2H_2	C_3H_6	C_2H_2	C_4H_8
2.6	8	52.6	6.8	0.8	12.0	0.1	1.7	25.6	0.4
	7	53.0	7.0	0.5	13.6	0.2	1.4	23.8	0.5
	6	53.3	7.1	0.5	12.6	0.1	1.3	24.4	0.7
2.0	8	56.0	6.5	0.5	9.7	0.1	1.2	25.6	0.4
	7	55.9	6.1	0.5	11.0	--	1.6	21.3	0.6
	6	56.8	5.8	0.7	9.4	0.3	1.3	24.8	0.9
	5	56.8	5.8	0.7	10.1	0.2	1.5	24.5	0.4
1.2	8	52.2	6.3	0.9	14.0	0.2	1.2	24.6	0.6
	7	53.6	6.1	0.6	12.4	0.1	2.0	24.8	0.3
	6	51.0	6.0	1.1	13.0	0.4	1.9	26.0	0.6
	5	51.0	6.1	0.9	13.1	0.2	2.0	25.8	0.9
0.6	8	52.2	5.9	0.7	13.0	0.1	2.0	25.8	0.4
	7	54.4	6.0	0.6	12.2	0.2	1.5	24.3	0.8
	6	55.1	5.9	0.7	11.8	0.2	1.4	24.0	0.9

TABLE 4. Mass Balance (per 1000 g Decomposed Feedstock) and Principal Power Indices for Codecomposition in Heavy and Light Petroleum Products

Amount decomposed, g		Amount obtained, g		Gas yield, liters/kg of feedstock	Yield of C_2H_2, liters/kg of feedstock	Yield of C_nH_{2n}, liters/m³ C_2H_2
heavy feedstock	light feedstock	gas	soot + losses			
Diesel oil 180.0	Petroleum product I 820.0	750-800	200-250	1380	358	540
Diesel oil 160.0	Petroleum product II 840.0	780-820	180-220	1420	360	620

Amount decomposed, g		Yield of H_2, liters/m³ C_2H_2	Yield of C_nH_{2n+2}, liters/m³ C_2H_2	Yield of soot, g/m³ C_2H_2	Feedstock consump., kg/m³ C_2H_2	Power consump., kWh/m³(NTP) C_2H_2
heavy feedstock	light feedstock					
Diesel oil 180.0	Petroleum product I 820.0	2000	308	558	2.79	11.0-12.0
Diesel oil 160.0	Petroleum product II 840.0	1980	330	500	2.70	11.0-12.0

TABLE 5. Compositions of the Gases Obtained by Codecomposition of Hydrocarbons

Initial petroleum product	Gas composition, vol.%							
	H_2	CH_4	C_2H_6	C_2H_4	C_2H_2	C_3H_6	C_2H_2	C_4H_8
n-Hexane	52.5	6.2	0.6	12.9	0.1	1.1	25.8	0.8
n-Heptane.	52.6	5.2	0.9	12.2	0.2	1.4	26.8	0.7
Hexene.	50.0	7.6	1.1	11.1	—	0.4	29.2	0.6
Cyclohexane	53.8	3.8	—	10.6	—	0.6	30.8	0.4
Methylcyclohexane	52.6	5.1	0.3	9.6	—	1.6	30.6	0.2
Benzene	59.0	3.7	—	4.0	—	0.3	33.0	—
Toluene	52.7	6.9	0.6	6.3	—	1.3	32.1	0.1

TABLE 6. Mass Balance (per 1000 g Decomposed Feedstock) and the Principal Power Indices for Codecomposition of Low-Boiling Hydrocarbons and Transformer Oil

Initial petroleum product	Amt.obtained, g		Yield					Feedstock consump., kg/m³ C_2H_2	Yield of soot, g/m³ C_2H_2	Power consump., kWh/m³ C_2H_2
	gas	soot + losses	gas, liters/kg feedstock	acetylene, liters/kg feedstock	olefins, liters/m³ C_2H_2	hydrogen, liters/m³ C_2H_2	unsat. comp., liters/m³ C_2H_2			
n-Hexane . . .	830	170	1390	362	480	2080	247	2.74	436	11.0
n-Heptane. . .	808	192	1295	348	504	1940	227	2.87	550	10.7
Hexene.	792	208	1200	352	378	1720	292	2.84	590	11.7
Cyclohexane .	758	242	1260	388	354	1750	116	2.60	625	11.0
Methylcyclo-hexane.	773	227	1220	374	379	1720	186	2.68	606	10.8
Benzene	475	525	840	278	117	1780	110	3.60	1900	10.1
Toluene	525	475	935	302	235	1640	220	3.30	1540	—

Considerable interest attaches to the power indices for the decomposition of petroleum products I and II and diesel oil, which were calculated from the corresponding mass balances (Table 4). The following yields (converted to 1 m³ acetylene) can be obtained: ~0.54-0.62 m³ lower olefins, ~2 m³ hydrogen, ~0.308-0.330 m³ unsaturated hydrocarbons, and ~500-558 g soot. The power consumption per m³(NTP) acetylene can be reduced by using more powerful and sophisticated power supplies.

We made a closer study of the character of the feedstock on the composition and yield of the cracking products for the following hydrocarbons: hexane, hexene, cyclohexane, benzene, heptane, methylcyclohexane, and toluene. These hydrocarbons served as the low-boiling feedstock, the heavy feedstock was transformer oil.

Tables 5 and 6 give the compositions of the gases formed by decomposition of the low-boiling C_5-C_7 hydrocarbons and the corresponding mass balances.

It will be seen from these tables that the maximum acetylene concentration (~32-33 vol.%) in the gas is observed during the decomposition of aromatic hydrocarbons, and the minimum concentration (~25-26 vol.%) in the case of paraffinic hydrocarbons. In the case of the naphthenic and olefinic hydrocarbons, the acetylene concentrations are between these values. However, it should be noted (cf. Table 6) that the absolute yield of acetylene reaches a maximum during the decomposition of paraffinic and naphthenic hydrocarbons (~348-388 liters/kg feedstock); the minimum amounts were obtained for aromatic compounds (~278-302 liters/kg of feedstock).

The concentration of the lower olefins in the gas and their absolute yield depend largely on the chemical nature of the feedstock; they are ~4-6.3 vol.% and ~117-235 liters/m³(NTP) C_2H_2, respectively, for the aromatic hydrocarbons, and ~12.0-14.0 vol.% and ~480-504 liters/m³(NTP) C_2H_2, respectively, for the paraffinic hydrocarbons.

SUMMARY

1. We have constructed and tested a laboratory apparatus for the codecomposition of a heavy and light feedstock in a high-voltage ac arc, designed for obtaining gas with high contents of acetylene and lower olefinic hydrocarbons.

2. It is shown that for the given types of feedstock (paraffinic hydrocarbons) the content of acetylene and olefinic hydrocarbons in the gas is largely independent of the interelectrode gap and the amount of low-boiling feedstock introduced into the apparatus per unit time. The mean acetylene content in the gas was 24-26 vol.%, that of the lower olefinic hydrocarbons approximately 8-14 vol.%.

3. It has been established that the formation of 1 m³(NTP) of acetylene requires the decomposition of ~2.7-2.8 kg of feedstock (~80% of light feedstock and ~20% of heavy). The power consumption in our case was ~11.0-12.0 kWh/m³(NTP) acetylene.

4. This method could be used for processing nonstandard gaseous gasolines and various heavy petroleum products which cannot be processed efficiently at present.

LITERATURE CITED

1.   Tatarinov, V. V., Russian Patent No. 40362 (1934).
2.   Pechuro, N. S., É. Ya. Grodzinskii, and O. Yu. Pesin, Symposium: Aspects of the Electrical Processing of Materials, No. 4, Izd. Akad. Nauk SSSR, Moscow (1962), p. 192.
3.   Pechuro, N. S., É. Ya. Grodzinskii, and O. Yu. Pesin, Gaz. Prom., No. 2: 47 (1963).
4.   Pechuro, N. S., and E. K. Starostin, Russian Patent No. 165709 (1964).

ON THE EFFECT OF THE ELECTRODE DIAMETER
AND CIRCULATION OF THE FEEDSTOCK ON THE DECOMPOSITION
OF LIQUID HYDROCARBONS IN ELECTRIC DISCHARGES

N. S. Pechuro, O. Yu. Pesin, and V. A. Filimonov

An analysis of reports on liquid hydrocarbon decomposition in arc discharges indicates that no study has been made of the combined effect of the electrode diameter and the amount of circulated feedstock on electro-cracking. Grodzinskii et al. [1], who used an ac high-voltage arc, established that circulation of the feedstock through the discharge zone is accompanied by an increase in its decomposition and the acetylene concentration in the gas, and by a fall in the power consumption per m^3(NTP) C_2H_2. In should be noted that in [1] the power of the arc discharge varied from ~7 to ~60 VA, but the electrode dimensions were constant.

The results of the present work have shown that, in addition to Q, the amount of circulated feedstock, the yield and composition of the gas are markedly influenced by the electrode diameters — the external diameter D, and the diameter d of the orifice in the high-voltage electrode (the feedstock enters the discharge zone via this hole).

The apparatus and procedure were described in [2]; however, in the present work the electrodes were positioned vertically in the reactor, not horizontally.

A petroleum product with initial boiling point 126°C and final boiling point 233°C was subjected to decomposition in an ac low-voltage arc with a high-voltage ignition.

The effect of the electrode diameters on electrocracking was studied both under stationary conditions and during circulation of the feedstock, the amount of the latter varying between 0 and 17.0 liters/min. The power of the arc discharge could be varied between about 0.4 and 4.0 kW by regulating I_{sc}, the short-circuit current of the power supply (in the range 10-50 A).

Table 1 shows the effect of the electrode's external diameter on electrocracking without circulation of the feedstock. It will be seen that an increase in this diameter from 10 to 40 mm reduces the gas yield. The gas has a relatively low acetylene content (18-23 vol.%) and a high concentration of hydrogen and methane. Note that the minimum acetylene content (about 18%) was obtained with large-diameter (40 mm) electrodes. This is evidently due to the fact that, in the absence of feedstock circulation, the gas formed (particularly in the case of large-diameter electrodes) does not have time to escape from the interelectrode gap during a discharge, and therefore undergoes secondary decomposition.

By high-speed filming of the decomposition of a liquid feedstock in an ac high-voltage arc, in a previous report [3] we established for the first time that secondary breakdown may occur in such a system.

The effect of the electrode diameter on electrocracking was also studied in the presence of feedstock circulation (Q = 5.5 liters/min).

TABLE 1. Variation of Gas Yield and Composition with External Diameter of the Electrodes (without Circulation of the Feedstock)

Ext. diam. of electrodes D, mm	I_{sc}, A	Gas yield V, liters/h	Content of the components, vol. %					
			H_2	CH_4	C_2H_6	C_2H_4	C_3H_6	C_2H_2
10	10	150	60.12	7.26	0.23	7.98	1.31	23.10
	20	282	62.38	6.96	0.28	7.28	0.65	22.45
	30	372	58.24	7.63	0.39	9.50	1.44	22.80
	50	530	58.70	7.67	0.35	9.44	1.19	22.65
20	10	120	64.56	4.82	—	7.51	0.81	22.30
	20	270	58.96	8.56	0.37	8.76	0.62	22.73
	30	320	60.10	7.60	0.26	9.00	1.02	22.02
	50	504	58.03	8.67	0.30	10.34	1.46	21.20
30	10	90	69.80	4.93	—	6.27	—	19.00
	20	240	59.18	8.52	0.31	9.52	1.17	21.30
	30	252	58.01	9.25	0.42	11.45	1.14	19.73
	50	450	54-55	8.43	0.52	11.79	2.45	22.26
40	10	51	67.36	5.77	—	7.09	0.83	18.95
	20	200	66.65	7.13	—	7.45	0.42	18.35
	30	240	64.34	7.52	—	9.08	0.64	18.42
	50	400	60.88	7.75	0.44	10.82	1.42	18.69

TABLE 2. Variation of Gas Yield and Composition with External Diameter of the Electrodes (Q = 5.5 liters/min, d = 5 mm)

Ext. diam. of electrodes D, mm	I_{sc}, A	Gas yield V, liters/h	Content of the components, vol. %					
			H_2	CH_4	C_2H_6	C_2H_4	C_3H_6	C_2H_2
10	10	258	61.14	3.89	0.25	5.10	0.22	29.40
	20	516	60.17	4.46	0.22	5.98	0.62	28.55
	30	714	60.93	4.69	0.20	5.99	0.64	27.55
	50	889	55.53	5.06	0.28	7.86	0.87	30.40
20	10	171	58.10	4.58	0.10	5.86	0.66	30.70
	20	438	60.00	5.20	0.19	6.25	0.76	27.60
	30	580	54.40	5.51	0.22	6.99	1.13	31.75
	50	870	52.28	7.64	0.30	8.65	1.58	29.55
30	10	114	58.40	5.13	0.22	6.70	0.45	29.10
	20	290	56.32	6.96	0.18	7.51	0.73	28.30
	30	430	55.08	6.80	0.19	8.06	1.07	28.80
	50	—	—	--	--	--	--	--
40	10	90	59.98	4.10	—	5.36	0.10	30.46
	20	190	53.73	7.44	0.23	7.68	0.57	30.35
	30	342	55.80	7.27	0.35	8.53	1.14	26.91
	50	510	55.48	8.28	0.34	9.02	1.38	25.50

The results (cf. Table 2) show that an increase in the electrode diameter is accompanied by a fall in the gas yield and the acetylene concentration. This fall is particularly marked when D = 40 mm and I_{sc} = 30 or 50 A. In the case of feedstock circulation we were able to increase the yield of the gas (in comparison with stationary conditions) and improve its composition.

It will be seen from Table 2 that in the case of 40-mm diameter electrodes a feedstock circulation of 5.5 liters/min is insufficient to ensure removal of the gaseous product from the discharge zone. In the case in question the gas yields differ little from those in Table 1 (with feedstock circulation), but the discrepancies are more pronounced for electrodes of smaller diameter.

TABLE 3. Variation of the Gas Yield and Composition with Amount of Circulating Feedstock (D = 30 mm, d = 16 mm)

Amt. of circulating feedstock Q, liters/min	I_{sc}, A	Gas yield V, liters/h	Content of the components, vol.%					
			H_2	CH_4	C_2H_6	C_2H_4	C_3H_6	C_2H_2
0.0	10	120	61.47	7.24	0.28	8.45	1.13	21.43
	20	198	58.30	8.41	0.33	8.69	1.56	22.71
	30	378	53.83	9.11	0.47	11.95	2.34	22.30
	50	550	57.64	8.69	0.47	10.41	1.98	20.81
3.0	10	140	61.13	5.13	—	4.89	0.45	28.40
	20	236	58.00	5.79	0.18	6.41	0.52	29.10
	30	510	57.45	5.50	0.31	6.72	0.82	29.20
	50	760	57.66	5.96	0.14	7.14	1.15	27.95
7.5	10	174	57.16	4.31	0.10	5.58	0.82	32.03
	20	282	57.57	4.83	0.24	5.94	0.84	30.58
	30	594	55.95	5.40	0.18	7.40	1.07	30.00
	50	900	55.00	6.39	0.15	8.30	1.16	23.00
13.5	10	224	58.28	4.04	0.28	6.02	0.82	30.56
	20	492	59.94	4.58	—	5.38	0.64	29.46
	30	690	59.78	4.26	0.13	5.91	0.85	29.07
	50	1150	56.77	4.64	0.17	6.45	1.06	30.91
17.0	10	252	59.28	5.42	—	4.77	0.27	30.26
	20	562	55.67	4.57	0.22	7.26	0.71	31.57
	30	844	54.14	5.19	0.33	6.99	1.33	32.02
	50	1230	53.92	5.05	0.29	7.13	1.41	32.20

TABLE 4. Variation of the Gas Yield and Composition with Amount of Circulating Feedstock (D = 10 mm, d = 5 mm)

Amt. of circulating feedstock Q, liters/min	I_{sc}, A	Gas yield V, liters/h	Content of the components, vol. %					
			H_2	CH_4	C_2H_4	C_2H_4	C_3H_6	C_2H_2
0.0	10	150	60.12	7.26	0.23	7.98	1.31	23.10
	20	282	62.38	6.96	0.28	7.28	0.65	22.45
	30	372	58.24	7.63	0.39	9.50	1.44	22.80
	50	530	58.70	7.67	0.35	9.44	1.19	22.65
5.5	10	258	61.14	3.89	0.25	5.10	0.22	29.40
	20	516	60.17	4.46	0.22	5.98	0.62	28.55
	30	714	60.93	4.69	0.20	5.99	0.64	27.55
	50	889	55.53	5.06	0.28	7.86	0.87	30.40
7.7	10	308	56.50	3.43	0.18	5.94	0.30	33.65
	20	550	55.00	3.23	0.15	7.22	1.00	33.40
	30	780	55.30	3.20	0.26	6.77	1.47	33.00
	50	1218	53.68	3.15	0.17	7.81	1.17	34.02
12.8	10	330	55.10	2.53	—	6.82	0.35	35.20
	20	570	51.87	3.50	0.23	7.70	0.60	36.10
	30	800	51.93	2.62	0.17	7.65	1.03	36.60
	50	1320	50.32	3.65	0.32	8.06	1.62	36.03
15.0	10	336	50.90	3.63	0.42	7.21	0.79	37.05
	20	660	50.60	3.50	0.40	7.82	1.68	36.00
	30	942	50.33	3.60	0.50	7.76	1.30	36.51
	50	1440	48.50	4.40	0.38	8.57	1.80	36.35

To make a closer study of the effect of the amount of circulating feedstock on electrocracking, we performed a series of experiments with a wide range of change of this parameter (up to 17 liters/min) and with D and d constant (30 and 16 mm, respectively). The results are given in Table 3. It will be readily seen that an increase in the amount of feedstock passing through the interelectrode gap is accompanied by a marked increase

TABLE 5. Variation of Gas Yield and Composition with Internal Diameter of the Electrode ($Q = 17$ liters/min, $D = 30$ mm)

Internal diameter of electrode d, mm	I_{sc}, A	Gas yield V, liters/h	Content of the components, vol.%					
			H_2	CH_4	C_2H_6	C_2H_4	C_3H_6	C_2H_2
5	10	150	57.01	3.53	0.22	8.10	0.33	30.81
	20	463	53.53	3.62	0.48	8.79	1.13	32.45
	30	740	51.30	4.45	0.50	9.15	2.10	32.50
	50	918	54.12	4.04	0.47	8.26	2.02	31.09
10	10	240	59.40	4.31	0.22	5.67	0.70	29.80
	20	510	54.49	3.71	0.36	7.61	1.23	32.60
	30	780	54.56	4.06	0.33	7.85	1.40	31.80
	50	1050	53.60	4.32	0.30	7.38	1.50	32.90
16	10	252	59.28	5.42	—	4.77	0.27	30.26
	20	562	55.67	4.57	0.22	7.26	0.71	31.57
	30	844	54.14	5.19	0.33	6.99	1.33	32.02
	50	1230	53.92	5.05	0.29	7.13	1.41	32.20
19.7	10	270	59.70	2.95	—	4.85	0.70	31.80
	20	636	56.17	4.19	0.17	6.77	0.86	31.84
	30	912	56.64	4.03	0.19	6.68	1.11	31.35
	50	1378	54.45	3.95	0.30	7.80	1.40	32.10

in the gas yield. With $Q = 17$ liters/min, the gas yield is more than double that observed under stationary conditions. There is a simultaneous increase in the acetylene concentration and a fall in the content of hydrogen, methane, and acetylene in the gas.

A similar series of experiments was performed with electrodes with $D = 10$ mm and $d = 5$ mm (Table 4). It will be seen that the relationships still have the same character. However, it must be emphasized that for the same values of Q and I_{sc}, the gas yield for small-diameter electrodes is greater than for large-diameter electrodes. This corroborates the laws previously established for the effect of D over a wide range of Q.

Furthermore, in the case of small-diameter electrodes ($D = 10$ mm, $d = 5$ mm) an increase in the amount of circulating feedstock is accompanied by a change in the composition of the gaseous product. With $Q = 15$ liters/min, the acetylene content reaches 36-37%, whereas in the case of electrodes with $D = 30$ mm it was less than approximately 32%.

The investigation showed that, in addition to D and Q, a marked effect on electrocracking is exerted by d, the diameter of the orifice in the high-voltage electrode (cf. Table 5). It will be seen that with Q and I_{sc} constant, an increase in d is accompanied by an increase in the gas yield. The gas composition remained practically constant under these experimental conditions.

It will therefore be seen that at a constant value of I_{sc} we can obtain different cracking-gas yields and compositions by changing the amount of circulating feedstock and the internal and external diameters of the electrodes.

High-speed filming showed that the arc discharge between circular electrodes can be struck in two ways (Fig. 1).

In the first case (a), the arc is at the outer edges of the electrodes and the gaseous products have free egress from the discharge zone. In the second case (b), the discharge channel appears within the interelectrode gap and is then displaced by the feedstock, together with the reaction products, toward the peripheries of the electrodes.

Case (b) is more typical for electrodes with large external and small internal diameter. Under these conditions the gaseous products may be decomposed by the next discharge, with an accompanying reduction in the acetylene concentration.

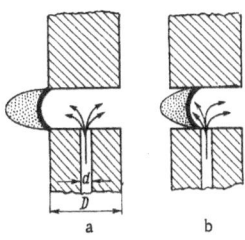

Fig. 1. Possible arc combustion variants.

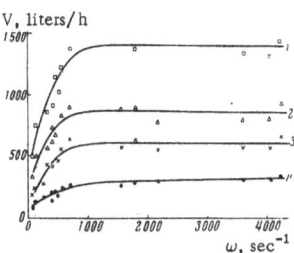

Fig. 2. Gas yield V versus nominal volumetric flow rate ω of the feedstock and short-circuit current I_{SC}. Current strength I_C (in A): 1) 50; 2) 30; 3) 20; 4) 10.

If we increase D and reduce d, we naturally reduce the residence time of the gaseous reaction products between the electrodes in the discharge channel. With an increase in the amount of circulating feedstock, the rate of displacement of the arc from the centers to the peripheries of the electrodes is heightened.

We cannot establish the true velocities of the feedstock and the decomposition products directly in the reaction zone, because there is no reliable method for determining the geometrical dimensions of the reaction zone, the temperature of the gaseous products, etc.

However, our investigations showed that the effect of such factors as D and d on electrocracking can be satisfactorily described by the parameter ω, which is the ratio of the amount of circulating feedstock to the volume of the interelectrode gap,

$$\omega = \frac{4Q}{\pi h\,(D^2 - d^2)} \text{ , sec}^{-1} \text{ ,} \qquad (1)$$

where Q is the amount of circulating feedstock in cm^3/sec, h is the interelectrode gap in cm, D is the external diameter of the electrode in cm, and d is the diameter of the electrode orifice in cm.

Since ω is expressed in sec^{-1}, we called it the nominal volumetric flow rate of the feedstock in the interelectrode gap. This parameter is not synonymous with the conventional volumetric flow rate because it does not allow for the amount of gas passing through the interlectrode gap (in addition to the feedstock), nor the degree of its thermal expansion at the time of formation.

From (1) we find that the nominal volumetric flow rate of the feedstock increases with the amount of circulating feedstock and the orifice diameter, and also increased with decreasing external diameter of the electrodes.

Since the same parameters (D, d, and Q) influence the gas yield, we can assume that the yield (V, liters/h) and the nominal volumetric flow rate of the feedstock must be related. In fact, the gas yields obtained at different values of D, d, and Q lie on the same $V-\omega$ curve (Fig. 2). It will be seen from this graph that an increase in ω is accompanied by an increase in the gas yield up to a certain limit. When $\omega > 1000$ sec^{-1}, the gas yield does not increase; at this flow rate the gaseous products are evidently removed from the interelectrode gap without undergoing secondary decomposition.

To verify that the (V, ω) curve obtained is correct, we performed a series of experiments with other values of D, d, and Q. The results are given in Table 6.

An analysis of Table 6 shows that the experimental gas yields differ very little from those in Fig. 2 for the corresponding values of ω and I_{SC}; this confirms that the above relationship is correct.

Oscillography of the process showed that an increase in the nominal volumetric flow rate of the feedstock is accompanied by a change in the electric parameters: the working potential and the mean discharge power increase; this may indicate arc stretching, i.e., an increase in the reaction space. For example, without circulation, and at low ω, the working potential is about 70-90 V, but at high ω it increases to about 250 V.

TABLE 6. Variation of Gas Yield and Composition and Mean Discharge Power with Nominal Volumetric Flow Rate of the Feedstock

Ext. diameter of electrodes D, mm	Int. diameter of electrodes d, mm	Amt. circulating feedstock Q, liters/min	Nominal volumetric flow rate ω, sec^{-1}	I_{sc}, A	Gas yield V, liters/h		Gas composition, vol.%						Mean discharge power N, VA	
					exptl.	graphical	H_2	CH_4	C_2H_4	C_2H_4	C_2H_4	C_2H_2	exptl.	graphical
20	10	5.48	388	10	206	190	60.38	4.00	--	5.20	0.49	29.93	638	647
				20	432	448	57.20	4.32	0.26	6.25	1.02	30.95	1207	1225
				30	630	660	58.30	4.10	0.25	5.80	1.15	30.40	1736	1780
				50	942	970	56.20	5.16	0.25	6.72	1.27	30.40	2820	2910
35	25	13.63	486	10	192	200	56.35	4.28	0.14	5.73	0.30	32.90	798	795
				20	468	480	55.50	4.08	0.27	6.18	0.77	33.20	1495	1430
				30	750	730	57.50	3.60	0.22	6.03	1.15	31.50	1956	1970
				50	1032	1050	57.40	3.03	0.22	5.95	1.20	31.30	3197	3200
15	8	13.63	1780	10	214	290	33.55	4.34	0.37	6.78	1.16	33.80	910	975
				20	617	610	54.60	3.64	0.38	6.59	1.09	33.70	1820	1800
				30	890	870	33.75	3.96	0.35	6.94	1.65	33.35	2630	2600
				50	1350	1380	34.30	4.80	0.30	7.60	1.00	32.00	3980	3910

TABLE 7. Variation of Power Consumption per m^3(NTP) C_2H_2 with Nominal Volumetric Flow Rate of the Feedstock

I_{sc}, A	Nominal volumetric flow rate of feedstock ω, sec^{-1}					
	0	100	450	2170	3620	4230
10	14.71	9.68	9.23	8.19	7.82	7.45
20	15.15	9.45	8.44	8.21	7.64	7.65
30	14.90	9.43	8.55	8.35	7.52	7.55
50	15.00	9.83	8.34	8.10	7.65	7.58

Figure 3 plots the mean discharge power versus the nominal volumetric flow rate of the feedstock. It will be seen that the curves are similar to those for the gas yields.

The data of Table 6 also confirm that the (N, ω) curve is correct.

From Figs. 2 and 3 we plotted the gas yield versus the discharge power (Fig. 4). It will be seen that the gas yield per unit consumed power (under the indicated experimental conditions) is constant and equal to the gradient of the straight line plotting V versus N.

Therefore, when a liquid hydrocarbon feedstock is decomposed in an ac low-voltage arc with ignition, the amount of gas formed is governed largely by the discharge power.

A change in the electrode diameters and the amount of circulating feedstock produces a different flow rate of the liquid and the decomposition products in the interelectrode gap, so that the arc power may vary. Furthermore, at a low feedstock flow rate there is increasing likelihood of secondary decomposition of the cracking gas, leading to a fall in the acetylene concentration.

One of the principal characteristics of the process is the power consumption per m^3(NTP) acetylene. Table 7 shows the change in this parameter with the nominal volumetric flow rate of the feedstock. It will be seen that the highest power consumption is observed in absence of feedstock circulation.

An increase in ω enables us to reduce the power consumption; under optimum conditions it is ~7.6 kWh per m^3(NTP) C_2H_2.

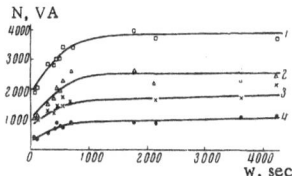

Fig. 3. Mean discharge power N versus nominal volumetric flow rate ω of the feedstock and short-circuit current I_{sc}. (For the meaning of the curve numbers, see Fig. 2.)

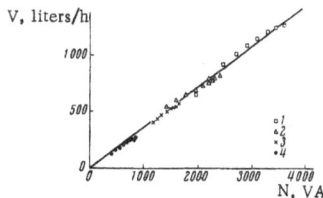

Fig. 4. Gas yield V versus mean discharge power N and short-circuit current I_{sc}. (For the meaning of the curve numbers, see Fig. 2.)

SUMMARY

1. We have established that the gas yield per unit time and the acetylene content in the gas can be increased by increasing the amount of circulating feedstock and the electrode's external diameter, and reducing its internal diameter.

2. It is shown that the effect of D, d, and Q reduces to a change in the nominal volumetric flow rate of the feedstock in the interelectrode gap, which determines the discharge power and the gas yield and composition.

3. An increase in the nominal volumetric flow rate enables us to reduce the specific power consumption per m^3(NTP) C_2H_2; under optimum conditions the specific power consumption is ~7.6 kWh at acetylene concentration ~36-37%.

LITERATURE CITED

1. Grodzinskii, É. Ya., N. S. Pechuro, and O. Yu. Pesin, Symposium: Arc Treatment of Metals, Izd. Akad. Nauk SSSR, Moscow (1963), p. 100.
2. Pechuro, N. S., O. Yu. Pesin, A. I. Gatarov, and M. K. Talibzhanov, Gaz. Prom., No. 9:40 (1964).
3. Pechuro, N. S., É. Ya. Grodzinskii, and O. Yu. Pesin, Symposium: Arc Treatment of Metals, Izd. Akad. Nauk SSSR, Moscow (1963), p. 96.

THE USE OF ELECTRICAL ENERGY FOR OBTAINING ACETYLENE FROM WET LIQUID HYDROCARBONS AND CRUDE PETROLEUM

N. V. Shishakov and F. M. Topol'skaya

Acetylene is produced industrially in four ways: the carbide method, high-temperature pyrolysis of the source material, and the thermooxidative (autothermal) and electrical methods.

Until very recently, the initial material for manufacturing acetylene in cracking processes has been gaseous hydrocarbons, principally methane. The use of liquid hydrocarbons for obtaining acetylene is only in the experimental stage.

Although methane is the cheapest material for obtaining acetylene, it is thermodynamically the least suitable, because its decomposition requires a much greater amount of heat than that for liquid hydrocarbons; furthermore, the acetylene content in the cracking gas as a whole is low owing to the inadequate conversion of methane to acetylene.

Interest attaches to the method of obtaining acetylene from liquid hydrocarbons in electric microdischarges (the Tatarinov method), which has been performed in the laboratory and in pilot plants, both in the USSR and abroad [1-7]. The advantages of this method are as follows: 1) it gives cracking gases with high acetylene concentrations; 2) a high yield of acetylene is obtained from the initial material; and, 3) organization of the process is simple and the specific energy costs suitably low.

However, this method, which is based on the use of anhydrous material (distillation residues, oils, gasoline, etc.), i.e., commercially refined petroleum products, is inevitably less advantageous from the economic aspect.

The use of naturally wet liquid hydrocarbons for cracking in electric microdischarges by the Tatarinov method or similar methods is complicated by the heterogeneity and demixing of the mixture of water and hydrocarbon; this prevents normal functioning of the process and its control, thus leading to production of gas with a variable composition. However, demixing of the mixture in the reaction zone can be eliminated by feeding to the latter a natural or man-made water—hydrocarbon emulsion which is sufficiently stable at the temperatures reached in the cracking process.

The use of water—hydrocarbon emulsions for the manufacture of acetylene by the electrical method enables us to use the cheapest feedstocks: liquid hydrocarbons, nonliquid refinery waste, and, most important of all, crude natural petroleum which has not been subjected to complete dehydration (in many cases even artificial emulsification is unnecessary, because a considerable part of the petroleum enters the tanks from the wells as natural stable emulsions).

Below we deal with the manufacture of acetylene by the use of water—hydrocarbon emulsions.

78

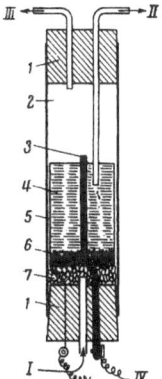

Fig. 1. Diagram of laboratory reactor.
1) Rubber corks; 2) quartz tube; 3) internal electrode; 4) emulsion; 5) outer electrode; 6) coke packing (intermediate catalyst); 7) paraffin pad. Flow lines: I) feed of emulsion; II) discharge of emulsion; III) gas outlet; IV) power supply for the reactor.

As the model emulsions we selected artificial emulsions with 17, 30, and 50% water.* In these experiments we used both "water-in-oil" and "oil-in-water" emulsions. The first of these types corresponded to natural stable water—petroleum emulsions, which are very difficult to process.

The laboratory apparatus for the investigations consisted of a reactor (cf. figure) and a power supply system, including devices for controlling the process and taking and analyzing samples and instruments for electrical measurements. The apparatus was connected to an ac lighting circuit working at 220 V.

The reactor was a transparent quartz tube of diameter 37 mm, with a rubber bung at each end. The electrodes were located in the lower part of the tube; the outer electrode was made of stainless steel, in the form of a perforated cylinder, tight against the wall of the tube. The other electrode, positioned along the tube's axis, was a metal rod of diameter 5 mm. The lower end of this electrode and the conductor sealed to the outer cylindrical electrode passed through the lower bung and were connected by terminals to the mains.

The feedstock (water, hydrocarbon, emulsion) was fed to the reactor via a tube in the lower bung, connected to the supply tank; it then passed through a layer of porcelain rings and packing (intermediate catalyst), flowed along a rubber tube to the receiver, and was then pumped bank to the supply tank. The liquid was thus recirculated.

The intermediate catalyst was a packing of electrode coke of 1-2 and 2-3 mm grain size. The height of the packing was 30-50 mm.

Mutual contact between the coke particles was prevented by the ascending feedstock, which separated them.

When the voltage was switched on, flashovers occurred between the particles and short-lived arcs were formed. Instantaneous explosive evaporation occurred at the discharge focus and the material was decomposed into highly superheated gaseous products, thus maintaining the bed of coke grains in a loose state. Formation of microdischarges throughout the bed and decomposition of the liquid passing between the particles of the bed could henceforth proceed for as long a period as desired, and the ascending current of liquid acted as a cooling agent.

The primary decomposition of the hydrocarbon to form acetylene can be schematically represented as follows:

$$C_m H_{2m+2} \rightarrow \frac{m}{2} C_2 H_2 + \left(\frac{m}{2} + 1\right) H_2.$$

The continuous circulation of cold liquid also ensured rapid removal of the cracking products from the reaction zone and restricted the development of secondary reactions leading to formation of ethylene, methane, and carbon; in the long run, this favored a high content of acetylene in the gas.

The flow rate of feedstock to the reactor was in the range of 0.5-1.5 liters/min, corresponding to a volumetric flow rate of 120-350 min⁻¹ and a 50-60°C temperature of the discharged liquid.

---

* These model emulsions were made under the supervision of L. L. Khotuntsev and L. S. Rapiovets.

The gases thus obtained were collected in the upper part of the reactor and fed to the analysis apparatus via the tube in the upper bung.

Before the work on water—hydrocarbon emulsions was begun, it was necessary to make absolutely sure that microdischarge effects could be obtained in a liquid hydrocarbon medium containing water with dissolved salts (a conductor of electricity).

According to Andrussow [4], water subjected to repeated decomposition in a field of microdischarges gave gases containing mainly hydrogen and carbon monoxide and a small amount of CO_2 and methane. However, his results were obtained for distilled water, i.e., a dielectric which can withstand the interparticle potential necessary for the formation of microarcs.

Separate experiments on the electrodecomposition of Moskvorets water (which has a distinctive natural hardness) showed that its dissolved salts do not prevent the formation of microdischarges. The experimental results thus confirmed that it was theoretically feasible to use wet liquid hydrocarbons for electrocracking.

The mean composition (11 analyses) of the gas obtained by electrodecomposition of water is as follows:

	From	To	Mean
CO_2	4.87	7.36	6.44
C_2H_2	0.29	0.64	0.50
C_nH_{2n}	0.17	0.43	0.25
O_2	0.85	1.26	0.90
CO	32.30	38.98	35.50
H_2	51.16	57.41	54.84
CH_4	1.39	2.90	2.17

The gas yield is 2.36 m^3(NTP)/g of decomposed water, and the mean consumption of the carbon in the packing 27.8 g/m^3(NTP) of gas.

The gas has the composition of a typical water gas with partial conversion of carbon monoxide to hydrogen and CO_2 (under these experimental conditions, conversion is due to the necessary temperature beyond the discharge zone and the catalyzing effect of the dispersed ash of the packing).

The ratio by weight of the oxygen and hydrogen in the gas is from 88:12 to 86:14, i.e., in all cases the gas has an oxygen deficit (in water, this ratio is 89:11).

The oxygen deficit in the gas can be attributed to the formation of formic acid in the reaction zones:

$$CO_2 + H_2 \rightarrow HCOOH,$$

The acid then dissolves in the water and is lost during gas analysis. This reaction was observed under analogous conditions by Eidus et al. [8].

On the other hand, the gas obtained by electrodecomposition of water always contains a small amount (up to 1% or more) of free oxygen, evidently due to formation of hydrogen peroxide in the discharge focus; it decomposes in the subsequent zones to form water and free oxygen.

In all the gas samples there is both qualitative and quantitative evidence of the presence of acetylene (fractions of a percent). Under these conditions acetylene may be formed by direct synthesis from the elements via C_2 radicals and atomic hydrogen at temperatures above 1700°C (as noted by Newland and Vogt [9]).

The possibility of acetylene formation from carbon monoxide and hydrogen by the reaction

$$2CO + 3H_2 = C_2H_2 + 2H_2O$$

was indicated by Berthelot.

TABLE 1. Composition of the Gases Obtained by Electrocracking of Gasoline (ibp 135°C) and a Petroleum Product with ibp 210°C

	Gasoline	Petroleum product
Gas composition, vol.%		
Acetylene	32.12	32.00
Ethylene	7.46	6.03
Hydrogen	54.01	57.46
Methane	6.41	4.51
Content in the gas, wt.%		
Carbon	81	82
Hydrogen.	19	18
Yield from converted feedstock, wt.%		
Acetylene	46	49
Acetylene−ethylene	57	60

Note: The gas yield is 1.25-1.35 m^3/kg. The gas was analyzed by chromatography; the unsaturated hydrocarbon group consists almost exclusively of ethylene; the methane hydrocarbon group contains up to 1 vol.% ethane. The carbon deficit in the gas (81-82% as against 86-87% in the feedstock) is due to loss with the soot.

Formation of acetylene in a medium containing water can also take place via carbides formed from metals in the ash of the coke packing, and from the electrode metals.

Table 1 gives the results of electrocracking of anhydrous gasoline (ibp 135°C) and a heavier petroleum product (aviation oil with ibp 210°C).

The gas composition is virtually the same as that noted in [1-4] for the electrocracking of anhydrous hydrocarbons.

At this juncture we must point out the advantages of this new method of manufacturing acetylene in comparison with conventional arc methods of methane cracking. The acetylene content in the gas obtained by the latter is less by a factor of 2-2.5, and the power consumption is 12.5 kWh/kg of acetylene.

The reason for the discrepancy between the electrocracking results for liquid and gaseous hydrocarbons must be sought in the difference between the media in which the discharge is effected, and in the character of the electric discharges. The method of cooling the decomposition products also has a marked effect on the acetylene yield.

In the arc method of methane decomposition the process is effected in the relatively small volume surrounding the discharge pinch, so that a considerable amount of the gaseous feedstock (methane) passes through the apparatus without decomposition and dilutes the gaseous products, thus reducing the acetylene concentration in the gas. In the new method, the whole volume of the reactor's discharge zone participates because the microdischarges penetrate the whole layer of the coke packing.

On the other hand, in the electrocracking of gaseous products the discharge is effected in the rarefied medium of the molecules of the incoming feedstock. This rarefaction is due not only to the character of the feedstock (gas), but also the high temperature within the apparatus. The number of feedstock molecules entering the discharge zone is therefore restricted. Therefore, to ensure the highest possible decomposition of the feedstock, we must make provision for the maximum discharge density throughout the whole space of the apparatus and use a feedstock of sufficient density. These requirements are met by the much recommended method of electrocracking in microdischarges in a bed of granular packing.

With increasing power consumption, the acetylene content in the gas as a whole displays little increase. This conclusion is not in line with the inferences of other investigators and is opposed to present hypotheses on

TABLE 2. Composition of the Gas Obtained by Electrocracking of Water—Hydrocarbon Emulsions

Petroleum product, emulsion type, and water content	Gas composition, vol.%						Decomposition, wt.%		Ratio of amounts of decomposed water and petroleum product
	CO_2	C_2H_2	C_2H_4	CO	H_2	CH_4	petroleum product	water	
Kerosene, "water-in-oil," 30%	1.06	28.63	6.94	7.14	50.70	5.53	89	11	0.123
Kerosene, "oil-in-water," 30%	0.72	25.30	6.72	10.42	50.50	6.34	86	14	0.162
Aviation oil, "water-in-oil," 30%	1.34	23.20	6.51	8.98	54.30	5.73	88	12	0.136
Petroleum, "water-in-oil," 17%	1.12	27.15	5.52	7.44	55.16	3.61	90	10	0.111

Note: The gas yield in terms of 1 kg decomposed feedstock is 1.62-1.63 m^3 for the water—kerosene emulsions and 1.71-1.73 m^3 for the other emulsions.

the state of the substance in the discharge zone, to the effect that the plasma formed by the discharge displays a constant composition (for the given substance), as do the end compounds combined from the plasma. An increase in the discharge power influences only the amount of substance converted to the plasma state, i.e., only the productivity of the cracking process in the discharge zone. However, this will only be correct when a change in the power is not accompanied by a change in the plasma's residence time in the discharge zone and the end products are retained in the zone for a constant period. We could not satisfy this requirement in our experiments because the buoyancy of the gas bubbles formed in the discharge zone (and therefore the rate of removal of the reaction products to the cold zone of the liquid) must increase with the power, thus preventing decomposition of the acetylene to hydrogen and soot.

Table 2 gives the results of the electrocracking of water—hydrocarbon emulsions. For the model feedstock we used the same products and also crude Novodmitrov petroleum with ibp 80°C and 29% yield of the fraction boiling in the range up to 200°C.

For emulsifiers we used petrolatum and an alkylphenylpolyhydroxyglycolic acid ester in a mixture with higher alcohols. The first of these was used for obtaining "water-in-oil" emulsions, the second for "oil-in-water" emulsions.

The electrocracking process was rapidly established and was just as stable as in the case of anhydrous hydrocarbons. Table 2 gives the composition of the gas obtained by electrocracking of water— hydrocarbon emulsions and the relative decomposition of the emulsions' components.

The maximum power was kept at 1000-1200 W, owing to the scale of the pilot plant.

The rate of decomposition of water in the electrocracking of emulsions was much lower than the cracking rate of the petroleum product. This is true for both types of emulsions and for all water contents; it may be attributed to the difference between the heats of decomposition of water and the liquid hydrocarbon. An increase in the emulsion's water content is accompanied by an increase in the relative fraction of water decomposed.

Our results show that the emulsion type has a marked effect on electrocracking. Thus (other conditions being equal), the use of a "water-in-oil" emulsion provides the best indices both as regards gas composition and the relative fraction of water decomposed.

For the same water contents of the emulsion (30%) and the same current powers, in the electrocracking of a "water-in-oil" kerosene emulsion the acetylene content in the gas is 28.6 vol.% and 0.123 kg of water is decomposed per kg of decomposed kerosene; for an "oil-in-water" emulsion, these figures are 25.3% and 0.162 kg, respectively.

TABLE 3. Mass Balance of the Electrocracking of Emulsions (in %)

	Petroleum product, emulsion type, and water content			
	Kerosene, "water-in-oil,"	Kerosene, "oil-in-water,"	Aviation oil, "water-in-oil,"	Petroleum, "water-in-oil,"
I. Carbon				
A. Of the petroleum product				
in the gas	61.4	57.1	45.6	55.0
in the soot	7.5	7.4	22.9	15.0
in the reaction with water	3.8	4.1	3.1	2.8
	72.7	68.6	71.6	72.8
B. Of the packing				
in the reaction with water	3.9	5.3	4.1	3.5
in the soot	1.0	0.8	1.0	1.3
	77.6	74.7	76.7	77.6
II. Hydrogen				
in the gas from the petroleum product	11.9	11.6	11.4	12.3
in water gas	1.2	1.4	1.3	1.1
	13.1	13.0	12.7	13.4
III. Oxygen				
from the water for the reaction with				
carbon	9.3	12.3	10.6	9.0
	100.0	100.0	100.0	100.0

Thus, the use of a "water-in-oil" emulsion for the manufacture of acetylene by electrocracking is economically advantageous. This is apparently due to the phenomenon noted by Ivanov [10], who studied the mechanism of emulsion drop combustion by high-speed photography. When a particle of a "water-in-oil" emulsion (i.e., a particle of water surrounded by a jacket of oil) is heated, owing to the fact that the boiling point of the water is much below that of kerosene, the water is converted to steam; being markedly superheated, it explodes and shatters the oil jacket into very fine particles which become intimately mixed with the water. It is very probable that this additional very fine dispersion of the petroleum product promotes its increased decomposition, both as a result of the electric forces of the discharge and the marked superheating.

The natural emulsions which are obtained during oil production and broken up at considerable cost during refining, are "water-in-oil" emulsions, i.e., the most suitable feedstock for electrocracking.

The best indices as regards gas composition and relative proportion of decomposed water relate to all cases of the electrocracking of monotypic water−kerosene, water− oil, and water−petroleum emulsions.

The formation of soot is an inevitable feature of acetylene manufacture by hydrocarbon cracking. It has been established that the yield of soot is directly proportional to the molecular weight of the emulsion's hydrocarbon component (Table 3). For example, the highest soot yield (22.9%) was obtained in the electrocracking of an aviation oil emulsion, i.e., the heaviest petroleum product. The soot yield falls to 15% when petroleum crude is used, and the lowest yield (7.4%) is obtained for kerosene.

Microscopic analysis of the solid cracking product showed that a certain amount (from 4 to 11%) of the coke packing's solid particles is intermixed with the soot during the process. The solid product can therefore not be utilized in the rubber industry. Under suitable production conditions, this product could be used for obtaining the coke for the intermediate catalyst.

Study of the mass balances of this process shows that the formation of water gas, which accompanies electrocracking of the hydrocarbon, is due to reaction of the water with the intermediate catalyst and the carbon of the petroleum product. The proportions of carbon (by weight) passing into the water gas from the coke and the hydrocarbon are very close. We have not determined whether the reaction of the petroleum product's carbon with water takes place via soot formation or reaction of the hydrocarbons with water vapor.

For "water-in-oil" emulsions, the consumption of the intermediate catalyst's carbon was virtually the same in all cases (3.5-4.1% in terms of the decomposed feedstock).

For petroleum crude containing 17% water, the power consumption per kg acetylene is 9-9.5 kWh, i.e., much less than in conventional methods for the manufacture of this gas. The power consumption is only 7.5 kWh in terms of acetylene and ethylene. *

SUMMARY

1. Cracking of dehydrated hydrocarbons (i.e., commercial, liquid hydrocarbons) in microdischarges for the manufacture of acetylene ensures maximum transformation of the feedstock to the target product (50% as against 25-30% for thermal methods of methane processing); the acetylene content in the end gas is 32 vol.% as against 10-17% in other methods.

2. It is shown that wet liquid hydrocarbons can be used for cracking in electric microdischarges. The gas thus obtained contains 32-35 vol.% acetylene and ethylene, i.e., more than twice as much as that obtained by thermal or thermooxidative cracking of methane and its homologs.

3. It is shown that acetylene can be manufactured from the cheapest hydrocarbon materials — watery emulsified waste products of petroleum refining and artificially emulsified natural petroleum which has been only partially dehydrated and desalinated.

4. After removal of acetylene and ethylene from the cracking gas the residual gas contains more than 65 vol.% hydrogen and CO and can be used as a feedstock for a number of chemical processes.

5. A preliminary technical—economic analysis shows that in petroleum refineries and petrochemical factories the cost price of acetylene manufactured by electrocracking in microdischarges will be 30% less than for any other method of manufacture, and the capital outlay will be no greater.

6. The energy costs for the artificial emulsification of the petroleum crude are very small and have little effect on the cost of acetylene manufacture.

LITERATURE CITED

1.   Tatarinov, V. V., Russian Patent No. 40362 (1934).
2.   Dobryanskii, A. F., and A. D. Kokurin, Zh. Prikl. Khim., 20:995 (1947).
3.   Kokurin, A. D., and V. V. Gruzdeva, Tr. Leningrad Tekhn. Inst. im Lensoveta, 57:111 (1959).
4.   Andrussow, L., Erdöl u. Kohle, 12:24 (1959).
5.   Cagas, F., M. Stand, and A. Lasarew, Erdöl u Kohle, 12:818 (1959).
6.   Obrezkov, V. D., Author's Abstract of Dissertation. Leningrad Tekhn. Inst. im Lensoveta (1963).
7.   Merkur'ev, A. N., Author's Abstract of Dissertation. MITKhT im. M. V. Lomonosova (1963).
8.   Eidus, Ya. T., et al., Dokl. Akad. Nauk SSSR, 27:4 (1940).
9.   Newland and R. Vogt, The Chemistry of Acetylene [Russian translation], IL, Moscow (1947).
10. Ivanov, V. M., Fuel Emulsions, Izd. Akad. Nauk SSSR, Moscow (1959).

---

* The net consumption after deducting the power costs for the reaction of water gas (6.8 kWh/m^3 CO + H$_2$).

ELECTROCRACKING OF SULFUROUS PRODUCTS
IN MULTIPLE ARC DISCHARGES

D. A. Sibarov and A. D. Kokurin

Virtually all petroleums contain a certain amount of sulfur, sometimes as much as 4-5%. Petroleums containing more than 1% sulfur comprise one third of the world's total production. In the USSR, the proportion of sulfur and high-sulfur petroleums is about 66% of the total production and the proportion of the latter is continuously increasing.

The diesel fuel components obtained from such petroleums have sulfur contents much higher than the values permitted by GOST. The residues left after the distillation of the light fractions also have high sulfur contents. For example, in the case of Chekmagush petroleum from the Carboniferous sequence, the residue after removal of the gasoline boiling below 200°C contains 3.62% sulfur [1].

The use of the residues from the distillation of high-sulfur petroleums as a boiler fuel is also restricted owing to corrosion of the equipment and to copious evolution of sulfurous gases.

The present paper studies the feasibility of using the distillation products of sulfur and high-sulfur petroleums for electrocracking in microdischarges.

In an earlier report [2] we showed that the electrocracking gas obtained by the decomposition of sulfurous petroleum products contains hydrogen sulfide, mercaptans, and carbon disulfide. Table 1 shows the sulfur content in the feedstock and its distribution in the decomposition products.

It will be seen from Table 1 that the cracking of sulfurous distillation residues is accompanied by a certain reduction in their sulfur content. Being less resistant to heat, the sulfur compounds evidently undergo greater changes than the hydrocarbon part. The free sulfur in benzene participates in the formation of the gas's sulfurous components. In the decomposition of a mixture of thiophene and benzene, or of a solution of sulfur in benzene, the sulfur in the gas was redistributed between hydrogen sulfide and carbon disulfide.

The study of the decomposition of sulfur compounds was continued on Arlan-Chekmagush and Shkapovo-Romashkino sulfur petroleums. Mixtures of the petroleums were rectified in a TsIATIM-58 apparatus. For electrocracking we selected diesel fractions which were removed at residual pressure 10 mm Hg and had boiling ranges of 200-320°C at normal pressure.

Decomposition was performed in a reactor of the type described in [3], but somewhat smaller and with current feed via the side wall.

In this work we used alternating current of potential 220 V at current strength 10-15 A.

The gas composition (cf. Table 2) was practically the same as in [2].

The quantitative composition of the sulfurous components in the electrocracking gas was established as follows. Hydrogen sulfide was determined iodimetrically; after the cadmium sulfide precipitate had reacted

TABLE 1. Sulfur Content in the Feedstock and Its Distribution in the Decomposition Products

Feedstock	Sulfur content, wt.%		Distribution of sulfur in the gas, %		
	in initial product	in residue	H_2S	CS_2	RSH
Distillation residues.........	3.35	3.14	29.6	45.7	24.7
Benzene + thiophene	3.0	3.0	65.0	35.0	—
Solution of sulfur in benzene ...	1.58	1.62	73.0	27.0	—

TABLE 2. Mean Composition of Gases Obtained in the Decomposition of Sulfur Petroleums

Petroleum	Gas composition, vol.%				Amount of gas obtained, liters
	C_2H_2	C_nH_{2n}	H_2	CH_4	
Arlan-Chekmagush	31.3	8.3	59.7	0.7	242
Shkapovo-Romashkino ..	31.4	8.0	60.0	0.6	210

TABLE 3. Mean Composition of the Sulfur Compounds in Electrocracking Gas

Fuel	H_2S, g/m^3(NTP)	RSH, g/m^3(NTP)	CS_2, g/m^3(NTP)	in feedstock, wt.%
Arlan-Chekmagush petroleum fraction	6.47	5.54	2.8	2.36
Shkapovo-Romashkino petroleum fraction	2.25	1.79	0.53	1.18

TABLE 4. Characteristics of the Feedstock and Electrocracking Products

Product	ρ^{20}_4, g/cm^3	ν_{20}, centistokes	S, wt.%	n^{20}D	Tar, wt.%
Shkapovo-Romashkino diesel fuel					
before cracking...............	0.8406	4.47	1.18	1.4695	—
after cracking................	0.8416	4.52	1.10	1.4711	—
after separation of the tars........	0.8391	4.40	1.02	1.4700	0.70
Soot extract					
with tars...................	0.8489	6.52	1.18	1.4738	—
after separation of the tars........	0.8467	5.76	1.17	1.4725	1.12
Arlan-Chekmagush diesel fuel					
before cracking...............	0.8568	5.74	2.36	1.4770	—
after cracking................	0.8606	5.87	2.25	1.4805	—
after separation of the tars........	0.8506	5.90	2.10	1.4750	1.0
Soot extract					
with tars...................	0.8674	7.54	2.36	1.4824	—
after separation of the tars........	0.8565	6.40	2.32	1.4784	1.91

TABLE 5. Ultimate Analysis of the Tars Obtained from Diesel Fuels

Feedstock	C	H	S	N + O
Shkapovo-Romashkino diesel fuel.....	78.93	8.93	5.22	6.92
Arlan-Chekmagush diesel fuel.......	79.35	8.03	5.93	6.69

TABLE 6. Group Composition of the Sulfur Compounds in the Feedstocks and Products

Product	RSSR, g/m^3(NTP)	RSR, g/m^3(NTP) aliphatic	aromatic	Residual sulfur, wt.%	S_{tot}, wt.%
Shkapovo-Romashkino petroleum					
before cracking.	0.04	0.20	0.49	0.4	1.18
	3.4	16.9	41.5	38.25	100.0
after cracking.	0.01	0.10	0.49	0.4	1.10
	0.9	9.1	44.6	43.68	100.0
Arlan-Romashkino petroleum					
before cracking.	—	0.18	0.61	1.5	2.36
	—	7.6	25.8	66.67	100.0
after cracking.	—	0.10	0.61	1.5	2.25
	—	4.4	27.1	68.44	100.0

TABLE 7. Characteristics of the Soot Obtained by Decomposition of Sulfur Petroleum Products

Feedstock	Soot, wt.% in decomp. feedstock	Volatile matter, wt.% according to GOST	by heating	S, wt.% initial soot	after removal of volatile matter, in accordance with GOST	after heating
Shkapovo-Romashkino Diesel fuel.	27.2	16.91	15.0	1.42	1.08	1.07
Arlan-Chekmagush Diesel fuel.	32.1	14.14	11.1	1.89	1.13	1.32

with iodine, the excess iodine was back-titrated with sodium thiosulfate. The mercaptans were combined in an alkaline solution of cadmium chloride and then determined iodimetrically after air had been passed through the solution to remove unsaturated gaseous compounds. The carbon disulfide content was determined by burning part of the gas, from which the other sulfurous components had been removed, in a quartz tube at 750°C; the sulfuric acid thus obtained was titrated with 0.02 N NaOH [4]. Flameless oxidation of the acetylene-rich gas was not obtained, so the amount of H_2SO_4 formed was checked gravimetrically from time to time. The air feed to the gas combustion tube was of a concentration below the mixture's explosive limit.

Table 3 shows the composition of the sulfur compounds in the gas.

It will be seen that for these fuels the contents of H_2S, RSH, and CS_2 in the gas depend on the feedstock's sulfur content. The figures for the mercaptan content are apparently much too high. It is probable that not all the unsaturated compounds were removed by the current of air and that some were sorbed on the cadmium hydroxide and reacted with the iodine.

After cracking had ended, the diesel fuel residue was discharged from the reactor and the soot filtered. The soot paste was extracted with a mixture of alcohol and benzene and the solvent distilled from the extract.

A weighed portion of the filtrate was covered with ASK silica gel for 10-12 h. The silica gel with the sorbed filtrate was then extracted with isooctane. The solvent was distilled from the extract and the filtrate freed from tar. The tar remaining on the silica gel was removed by extraction with an alcohol—benzene mixture. We thus obtained a tar-free soot extract and the tar of the extract.

The filtrates and extracts obtained before and after removal of the tar were analyzed for the sulfur content and their densities, kinematic viscosities, and refractive indices determined. The results are shown in Table 4.

It will be seen from Table 4 that cracking is followed by an increase in the density, viscosity, and refractive index of the liquid residue, and a fall in the sulfur content of both fuels by 4.5%.

Thus, as in our previous work, there is a rather marked decomposition of organic sulfur compounds in the arc discharges. The increase in the above-mentioned physical characteristics of the filtrates depends mainly on the tars formed during electrocracking. After removal of the tars, these indices fall (the S content in the Shkapovo-Romashkino fraction decreases by 7%, that in the Arlan-Chekmagush fraction by 6.7%). For each type of fuel the tar content in the extract is higher than in the filtrate, evidently because the soot retains a greater amount of tarry substances owing to its marked absorption capacity. Table 5 shows the ultimate analysis of the tars from the cracked fuels.

It will be seen from Table 4 that there is a marked difference between the soot extracts before and after tar removal; the sulfur content in the extracts was the same as in the uncracked fuels.

In the decomposition of sulfurous products in discharges, interest was attached to the change in the sulfur compounds in the fuels and in the cracking residues. It is known that the various groups of sulfur compounds in petroleums undergo marked changes at high temperatures, which become increasingly pronounced with the temperature [5, 6].

The group composition of the sulfur compounds was determined by the Faragher—Ball method [7, 8]. Although this method has shortcomings, such an analysis gives us an idea of the group distribution of the sulfur compounds.

The total sulfur contents of the fuels were determined by the lamp method in accordance with GOST 1771-48. The mercaptans were determined by potentiometric titration, using a silver sulfide electrode [9]. The disulfides were reduced to mercaptans, which were determined as above. The aliphatic sulfides were determined by means of mercurous nitrate. The aromatic sulfides and thiophenes were determined by treating the fuels with mercury nitrate. The sulfur compounds remaining in the products were classed as "residual sulfur."

The results are given in Table 6. The upper rows of figures indicate the absolute content of the given group of sulfur compounds in the fuel, the lower rows the percentage content of each group. It will be seen from Table 6 that electrocracking is accompanied by a change in all the groups of sulfur compounds. The disulfide content in the fuel obtained from Shkapovo-Romashkino petroleum was reduced to nearly one quarter of the initial value. There was a marked fall in the aliphatic sulfide contents of both fuels, and a slight increase in the aromatic sulfide and "residual sulfur" contents. The elementary sulfur (1.8%) observed in the liquid residue after the cracking of the diesel fuel of Shkapovo-Romashkino petroleum is not really free sulfur. It would appear that mercury metal may extract the unsaturated sulfur compounds, which can be formed during electrocracking in arc discharges [10].

In our previous reports [11, 12] we indicated the presence of small amounts of unsaturated compounds in the liquid residue after cracking.

We also obtained the infrared spectra of the feedstocks and the liquid products. The spectra of the cracking residues contained a band at 3272 cm^{-1}; the same band has been observed in the spectra of octane after cracking. According to Bellamy [13], this band belongs to valence vibrations of the hydrogen atom in monosubstituted acetylenes.

The group analyses of the diesel fractions (cf. Table 6) show that in both cases the bulk of all the sulfur compounds are included in the residual and sulfide sulfur.

The electrocracking of sulfur petroleum products was accompanied by considerable soot formation. After extraction with a mixture of benzene and alcohol, the soot was dried at 105°C. We determined its sulfur content, volatile matter content (in accordance with GOST 628-41), and weight loss after heating at 820°C for 6 h in oxygen-free nitrogen. The results are given in Table 7.

The difference between the volatile matter yields determined in accordance with GOST and by heating is due to the different rates of heating of the soot samples. After removal of the volatile matter, the sulfur content in the soot decreases somewhat, but is still considerable.

When the soot was roasted, the volatile products were passed through absorption flasks containing a cadmium chloride solution and an alcoholic solution of diethylamine. Sulfur was removed from the soot mainly as H_2S, and only traces of carbon disulfide were found.

SUMMARY

1. The gas obtained by electrocracking of sulfur petroleum products contains hydrogen sulfide, mercaptans, and carbon disulfide, the amounts of these products depending on the sulfur content in the initial substance.

2. It has been established that cracking is accompanied by a reduction in the sulfur content of the liquid product and an increase in the density, viscosity, and refractive index. When the tar is removed from the liquid residue, these indices decrease.

3. It is shown that the soot formed contains considerable amounts of sulfur; this is not completely removed by roasting for 6 h in nitrogen.

4. Study of the group composition of organic sulfur compounds has shown that the sulfur undergoes changes during electrocracking, leading to the formation of compounds with higher resistances to heat in the liquid residue.

LITERATURE CITED

1. Ivchenko, E. G., and G. V. Sevast'yanova, Sulfur and High-Sulfur Compounds of Bashkirian Petroleum, Gostoptekhizdat, Moscow (1963).
2. Kokurin, A. D., V. D. Obrezkov, and D. A. Sibarov, Zh. Prikl. Khim., 36:424 (1963).
3. Kokurin, A. D., and V. D. Obrezkov, Zh. Prikl. Khim., 25:2574 (1962).
4. Rapoport, F. M., Tr. GIAP, 35:275 (1952-1953).
5. Obolentsev, R. M., and A. V. Mashkina, Hydrogenolysis of Organic Sulfur Compounds of Petroleum, Gostoptekhizdat, Moscow (1961).
6. Skripnik, E. I., The Chemistry of the Organic Sulfur Compounds Present in Petroleum and Petroleum Products. Izd. Bash. Fil. Akad. Nauk SSSR, Ufa (1958), p. 43. [English translation: Israel Program for Scientific Translation (1963). Available from office of Technical Services, U. S. Department of Commerce, Washington, D. C.].
7. Faragher, W. F., and J. C. Morrel, Ind. Eng. Chem., 10, 1091 (1907).
8. Ball, J. S., U. S. Bureau of Mines Report of Investigation 3591 (1941).
9. Methods for the Analysis of Organic Compounds of Petroleum, their Mixtures and Derivatives, Izd. Akad. Nauk SSSR, Moscow (1960).
10. Cherkov, Ya. B., and V. N. Zrelov, Zh. Prikl. Khim., Vol. 31 (1958).
11. Kokurin, A. D., O. N. Setkina, and V. V. Gruzdeva, Tr. Leningrad Tekhn. Inst. im Lensoveta, 51:102 (1959).
12. Kokurin, A. D., and V. V. Gruzdeva, Tr. Leningrad Tekhn. Inst. im Lensoveta, 51:113 (1959).
13. Bellamy, L., The Infrared Spectra of Complex Molecules [Russian translation], Moscow-Leningrad, IL (1963).

THE EFFECT OF CERTAIN FACTORS ON ELECTROCRACKING
IN MICRODISCHARGES

A. D. Kokurin, E. A. Kolodin, and V. D. Obrezkov

Study of electrocracking in microarc discharges has established that the composition of the decomposition products, the output of the reactor, and the consumption of power and electrode carbon depend markedly on the reactor's design, the number and size of the carbon granules ("mobile electrodes"), the fixed-electrode gaps, the circulation of the liquid, the current potential and strength, and the pressure in the system.

The experiments were performed in cylindrical and rectangular reactors. In cylindrical reactors with two or more conducting electrodes, it was difficult to ensure uniform gas formation throughout the whole mobile-electrode volume at 127 and 220 V. Furthermore, the specific productivities of cylindrical reactors with vertical electrodes were less than those of rectangular reactors with horizontal electrodes [1]. Most of the work was therefore carried out in rectangular electrodes with two or more conducting electrodes.

The laboratory-type reactors with gas outputs of 0.1-1 m^3/h had two or three conducting electrodes of length 100-150 mm. The pilot-type reactor with gas output up to 10 m^3/h had four conducting electrodes of length up to 500 mm. Figures 1 and 2 give diagrams of these reactors.

In the laboratory the experiments were performed with and without circulation of the feedstock, and also with mechanical mixing of the mobile electrode carbon [1, 2]. Such mechanical mixing by a stirrer in the reactor did not afford appreciable advantages in comparison with circulation of the feedstock and was therefore stopped.

Figures 3 and 4 give diagrams of the two types of apparatus.

It was established by numerous experiments that A, the optimum weight of the mobile electrode carbon for the highest output, maximum acetylene content in the gas, and minimum consumption of power and electrode carbon, is a variable which depends on the electric current parameters U, the fixed electrode gap l, the size of the mobile electrode—carbon granules C, and the circulation of the liquid K, i.e., $A = f(U, l, C, K)$.

In experiments with kerosene at potential 220 V, fixed-electrode gap 50 mm, and mobile electrode—carbon granules of size 8-10 mm, the gas yield varied with the height of the bed of electrode carbon, as follows:

Height of layer, mm ..	10	15	25	40
Gas yield, m^3/h	0.092	0.394	0.648	0.540

In this case the optimum specific load of mobile electrode carbon (in grams per cm^2 of the reactor's effective area, i.e., the area between the stationary electrodes) was 2-3 g/cm^2.

With potential 380 V, stationary electrode gap 40-50 mm, and mobile electrode carbon granules of size 20-30 mm, the optimum specific load increased to 7-9 g/cm^2.

90

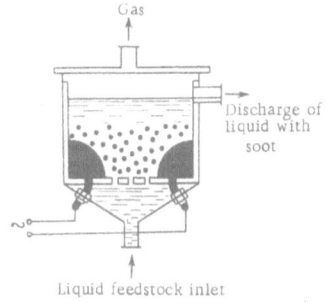

Fig. 1. Laboratory reactor.

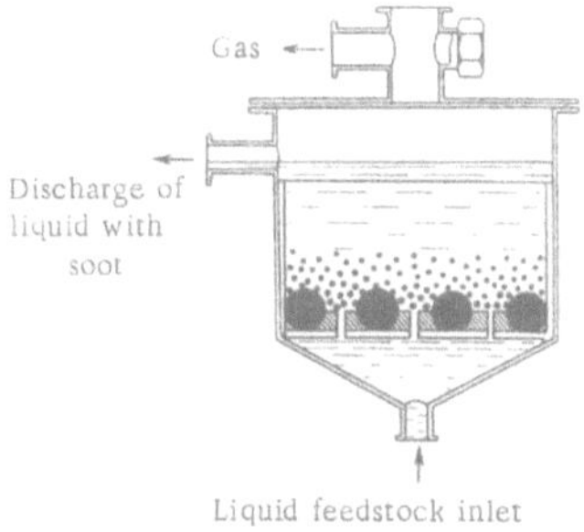

Fig. 2. Pilot reactor.

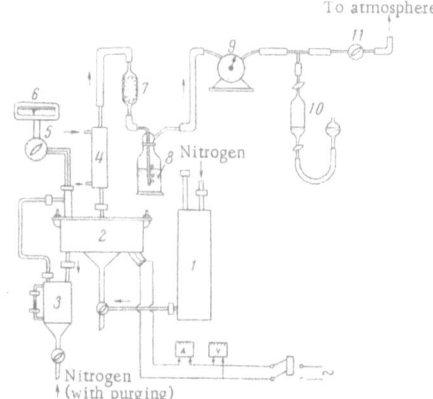

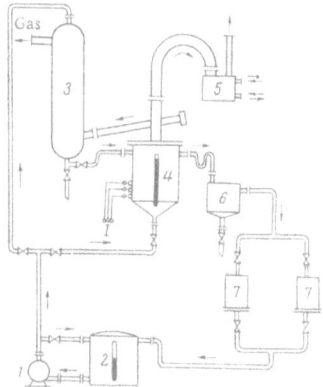

Fig. 3. Diagram of laboratory apparatus. 1) Supply tank; 2) reactor; 3) discharge tank; 4) condenser; 5) switch; 6) pyrometer; 7) wool filter; 8) bubbler soot trap; 9) gas meter; 10) aspirator; 11) tap for discharging gas to atmosphere.

Fig. 4. Diagram of the pilot plant. 1) Pump; 2) supply tank; 3) scrubber; 4) reactor; 5) water seal; 6) settling tank; 7) filter.

A reduction in the specific load is accompanied by a decrease in the reactor's productivity, but the gas composition and the power consumption are unchanged.

If the specific load is increased above these limits, the reactor's productivity falls, the power consumption increases, and the gas composition is changed. This is due to the fact that the weight of the overlying mobile carbon granules reduces the mobility of the layers beneath them, the current between the feed electrodes is short-circuited, and power losses due to Joule heat increase. This effect is not completely eliminated by circulation of the liquid feedstock. With carbon granules smaller than 10 mm and circulation of the feedstock, the optimum charge can be increased by a factor of 1.2-1.5 and the reactor's productivity somewhat increased.

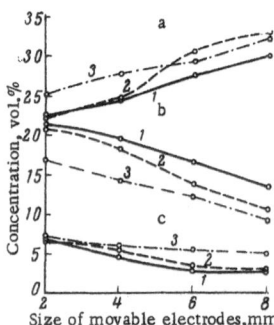

Fig. 5. Composition of the hydrocarbon components of the gas, plotted versus size of the mobile electrode carbon granules and the character of the feedstock. 1) n-Octane; 2) kerosene; 3) shale tar. a) Content of acetylene; b) content of olefinic hydrocarbons; c) content of paraffinic hydrocarbons.

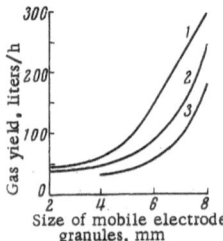

Fig. 6. Effect of size of the mobile electrode carbon granules and character of the feedstock on the gas yield. 1) n-Octane; 2) shale tar; 3) benzene.

TABLE 1. Change in Gas Composition with the Stationary Electrode Gap and the Size of Mobile Carbon Granules

Fixed-electrode gap, mm	Size of movable carbons, mm	C_2H_2 content in the gas, vol.%
50	6-7	25.2
40	6-7	29.3
25	6-7	33.0
40	8-10	33.4

Owing to the dependence of the specific charge of mobile electrode carbon granules on so many factors, their optimum number in the arc discharge zone must be determined by experiment for each case.

Figure 5 plots the composition of the gas's hydrocarbon components versus the size of the mobile electrode carbon granules. The experiments were performed at fixed-electrode gap 50 mm, working length of the fixed electrodes 100 m, potential 220 V, and weight of the mobile carbon 100 g.

A decrease in the size of the mobile carbon granules while the other conditions remain unchanged is accompanied by a reduction in the gas's acetylene content and an increase in the content of the olefinic and paraffinic hydrocarbons.

Owing to the high mobility of small carbon granules, low-power arcs are struck between them, the combustion time of which is shorter. In consequence, decomposition of part of the feedstock ceases abruptly in the intermediate stages and a small amount of high-molecular compounds is formed owing to partial polymerization.

However, the experiments showed that when 10- to 40-mm carbon granules are used under these conditions, the gas composition was virtually unchanged [3]. Therefore, the combustion time, and the volume and power of the arc obtained with 8- to 10-mm carbon granules ensure total decomposition of the substances fed to the arc discharge without formation of high-molecular liquid products.

A decrease in the size of the carbon granules also affects the reactor's productivity and the power consumption.

Figure 6 plots the gas yield versus the size of the mobile "electrodes." It will be seen that with decreasing granule size the reactor's output falls sharply.

In addition we found how the nature of the initial feedstock affects the gas output. In electrocracking of aromatized feedstock (benzene, shale tar) we get a lower yield of acetylene and total gas per unit of feedstock than in the decomposition of paraffinic and naphthenic hydrocarbons.

The experimental data show that the reactor's output and the gas composition depend on the number and size of the mobile electrode granules and the character of the feedstock.

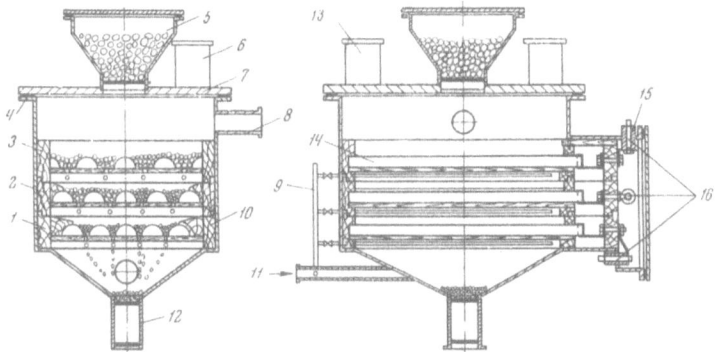

Fig. 7. Diagram of multistage reactor for electrocracking of liquid products. 1) Frame of lower stage; 2) frame of middle stage; 3) frame of upper stage; 4) gasket; 5) loading hopper for the mobile electrodes; 6) gas-sampling pipe; 7) cover; 8) discharge connecting pipe; 9) feed pipe for feedstock beneath the frame; 10) body; 11) connecting pipe for feeding the stock to the reactor; 12) sluice valve; 13) loading pipe for viscous products; 14) fixed electrode; 15) insulator; 16) current lead.

It was also established that a change in the fixed-electrode gap and the size of the mobile granules is accompanied by a change in the gas composition and the reactor's output. The results of these investigations are given in Table 1.

It will be seen that for 6- to 7-mm granules, a reduction in the fixed electrode gap from 50 to 25 mm is accompanied by an increase in the gas's acetylene content from 25 to 33%. At potential 220 V the reactor's productivity (V, m^3/h) changes with the stationary electrode gap (l, mm) and the movable electrode diameter (d, mm) as follows:

l/d	5.86	3.14	2.85	2.4	2.0	1.75	1.28
V, m^3/h	0.954	2.0	2.8	4.98	6.0	6.84	2.04

At 220 V, a higher productivity was obtained with $l/d \simeq 1.7$-2.5. At 380 V, the optimum ratio is 2-3. In this range the gas composition and the acetylene content remain virtually constant.

As a result of these investigations we designed the multistage reactor shown in Fig. 7.

To keep the output, gas consumption and power consumption constant, we must keep the ratio l/d and the specific carbon load within the optimum range in each frame.

The size of the mobile granules in each frame is determined by the width of the slot in the bottom of the frame. The presence of the necessary number of granules in each frame is ensured by adding them to the upper rack. During the operation of the reactor, they fall through the slot and reach the appropriate frame. This type of reactor design also provides more efficient utilization of the electrode—carbon granules.

The circulation of the liquid feedstock ensures removal of excess crushed carbon from the lower frame and of the soot from the reactor, cooling of the reaction space, and greater stability of the process. The circulation velocity of the liquid is governed by the size of the carbon granules and the reactor's output.

The consumption of the circulating feedstock is between 0.02 and 0.2 m^3/m^3 of gas.

It was found that under optimal conditions the power consumption for the laboratory apparatus was 10-12 kWh/kg C_2H_2, and that of the pilot apparatus 9.2 kWh/kg C_2H_2.

At potential 380 V, the power consumption for the pilot plant fell to 8 kWh/kg C_2H_2.

The power consumption for decomposition of the various organic liquid products to acetylene and hydrogen was approximately 2.5-3.5 kWh/kg C_2H_2 [1]. The practical consumption for the electrocracking of liquid products, determined from the material and fuel balances, was about 5.8-6 kWh/kg C_2H_2.

We could thus make a further reduction in the power consumption by improving the process and reducing the Joule heat losses.

SUMMARY

1. The optimum weight and specific consumption of mobile electrode carbon, the maximum acetylene content, and the minimum power consumption depend on the discharge parameters, the fixed-electrode gap, the size of the mobile electrode carbon granules, and the circulation of the liquid products.

2. An increase in the specific electrode—carbon load above the optimum is accompanied by a reduction in the reactor's output, an increase in the power consumption, and a change in gas composition.

3. A reduction in the size of the mobile carbon granules while the other conditions are unchanged is accompanied by a reduction in the gas's acetylene content and an increase in its content of olefinic and paraffinic hydrocarbons.

4. The reactor's output and the gas composition depend on the character of the feedstock. In the decomposition of aromatized products a smaller amount of gas and acetylene per unit of the feedstock is formed than in the case of paraffinic and naphthenic hydrocarbons.

LITERATURE CITED

1.   Kokurin, A. D., and V. D. Obrezkov, Tr. Vses. Nauchno-Issled. Inst. Topliva, No. 11:107 (1962).
2.   Kokurin, A. D., V. D. Obrezkov, and É. A. Kolodin, Zh. Prikl. Khim., 35:1379 (1962).
3.   Kokurin, A. D., É. A. Kolodin, et al., Tr. Vses. Nauchno-Issled. Inst. Topliva, No. 13:45 (1964).

THERMAL DECOMPOSITION OF LIQUID PETROLEUM PRODUCTS
MIXED WITH SOOT FORMED DURING ELECTROCRACKING

N. S. Pechuro, E. Yu. Bulychev, and A. N. Merkur'ev

When liquid hydrocarbons are decomposed in an electrical discharge, as well as gas containing up to 30 vol.% of acetylene, we also obtain condensation products (soot) which we find in a dispersed state in the processed products. It has been shown [1-3] that the yield of soot depends largely on the nature of the initial raw material, and increases as we go from volatile to high-boiling petroleum products. It has been found experimentally [4] that the presence of small quantities of condensation products (15-20 g/liter) in the circulating liquid feedstock, when high-voltage discharges are being used, promotes the decomposition process by facilitating electrical breakdown; but larger concentrations (greater than 50 g/liter) cause short-circuiting of the electrodes by soot "bridges." In a continuous process it is therefore necessary to remove part of the liquid product, together with the soot, from the cycle.

The soot can be removed (when working with volatile petroleum products) by filtration [5] or centrifuging. However, these methods (in the absence of preliminary dilution with volatile products) would be difficult to apply to viscous high-boiling types of feedstock. For this reason we must try to find appropriate methods for processing the latter.

From our viewpoint, possible uses of the processed feedstock removed from the cycle might be thermal treatment in metal stills and specially constructed plants of the coke-oven type [6-8], in rotating-drum furnaces used for sludge disintegration [9], or in apparatus with mobile packings [10-15]. In all these cases, together with the condensate we get an ashless solid residue and a gas (in varying proportions, depending on the temperatures used).

In this article we shall give our results on the thermal decomposition of processed liquid electrocracking products by semi-coking (i.e., low-temperature carbonization) and on mobile packings.

As we have stated, the raw material for further thermal processing was liquid products containing dispersed soot, removed from the electrocracking cycle. We shall denote these substances by the following abbreviations: processed products from electrocracking of crude petroleum, PP_c; of Diesel oil, PP_{do}; of vacuum gas-oil, PP_{vg}. Table 1 gives some properties of these samples.

We used a 200-g aluminum retort in which the temperature was gradually raised to 500°C over 3 h. In all the experiments the retort was loaded with about 200 g of raw material. The results are given in Table 2. It will be seen that, for semicoking in an aluminum retort, the main process is distillation of liquid products, and there is little thermal decomposition. This is shown by the small yield of gas (about 2-3%) and the high yield of liquid products (87-90%). The sulfur compounds in the gas consist mainly of hydrogen sulfide (38-40 g/m³*) and a relatively small amount of organic sulfur compounds (about 0.8-40 g/m³). In developing a

---

* All gas volumes are at NTP — Translator.

TABLE 1. Characteristics of Raw Materials for Thermal Treatment

	PP_c	PP_{do}	PP_{vg}
Total sulfur, wt.%	1.67	1.33	0.83
Soot content, wt.%	6.10	5.50	5.60
Content of liquid products, wt.%	93.90	94.50	94.40

Note: The soot content was found by extraction in a Soxhlet apparatus.

TABLE 2. Product Yields and Gas Composition from Semicoking in Aluminum Retort

	Type of raw material		
	PP_c	PP_{do}	PP_{vg}
Yield of semicoking products, wt. %			
condensate	85.4	92.4	91.1
semicoke.	12.9	5.6	6.1
gas + losses	1.7	2.0	2.8
Total	100.0	100.0	100.0
Gas composition, %			
H_2	9.4	7.6	11.4
CH_4	38.9	23.8	40.5
C_2H_6	20.6	18.8	22.7
C_2H_4	2.9	19.8	4.7
C_3H_8	11.6	11.4	8.7
C_3H_6	5.0	10.9	7.2
C_4H_{10}	4.4	5.3	2.6
ΣC_4H_8	7.2	2.4	2.2
Total	100.0	100.0	100.0
Sulfur compounds in gas, g/m^3:			
H_2S	48.0	43.0	38.0
organic sulfur compounds (sum)	1.12	1.04	0.80

system for electrocracking with a semicoking unit included in the cycle, the liquid distillate products may be subjected to further decomposition.

The second stage of the investigation was performed with a larger laboratory plant, of which Fig. 1 gives a schematic diagram. Before the start of an experiment, the initial product (400-500 g) was mixed with 15-20 times its weight of cast-iron balls (d ≈ 2 mm), and the mixture was placed in the receiving bunker 1. The latter was made airtight by means of a lid with a hydraulic seal 2. The receiving bunker was connected to the bottom discharge unit by a seamless metal tube (d = 80 mm, l = 2100 mm), which acted as the reaction zone inside which the packing moved and the products were decomposed. This zone was heated by electric furnaces 3, and its temperature was controlled by thermocouples 4 at various levels. The required rate of motion of the packing was achieved by means of a special discharge mechanism 5. The packing discharged from the cycle was dropped into a bottom bunker 6. The discharge mechanism was constructed so as to permit screening off of the solid decomposition products into a special bunker 7. The vapor— gas mixture obtained from thermal decomposition was led off via a tube 8 and then passed in sequence through a condenser 9, a receiver 10, a trap 11,

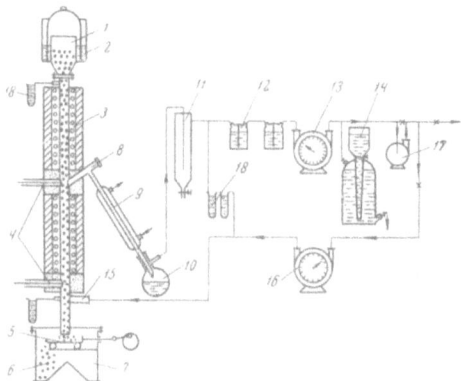

Fig. 1. Diagram of apparatus.

TABLE 3. Yields of Products from Thermal Decomposition of Initial Specimens on Mobile Packings at Various Temperatures

Initial petroleum product	Temp., °C	Yields of products, wt.%			
		gas	volatile fractions	liq. product (condensate)	solid residue + losses
PP$_c$	500	19.9	2.7	62.7	14.7
	600	31.5	2.5	38.0	28.0
	700	35.6	3.2	31.8	29.4
	800	38.1	4.4	26.4	31.1
PP$_{do}$	500	14.7	2.4	62.4	20.5
	600	31.0	1.7	42.0	25.3
	700	34.9	4.3	34.9	25.9
	800	36.1	6.1	26.2	31.6
PP$_{vg}$	500	12.3	3.1	73.0	11.6
	600	26.8	4.1	54.1	15.0
	700	31.8	4.5	46.0	17.7
	800	40.0	5.6	28.6	25.8

and oil absorbers 12 to trap the volatile fractions. The purified gas passed through a gas meter 13 and was discharged from the cycle. An average sample of the gas was taken in gasometer 14. To prevent vapor—gas mixture from reaching the lower bunkers, arrangements were made for blowing it out by leading part of the purified gas back into the system via a branch pipe 15. About 10-15% of the gas was recirculated; the amount was measured by gas meter 16. Gas was drawn off and circulated by blower 17. Some excess pressure was maintained in the system, and was measured by U-tube manometers 18.

The mass balances of the process were compiled from measurements of the weights of initial product fed in, output of volatile fraction (from decomposition of condensate) absorbed in the oil traps, and gas. The solid residue and losses were determined by subtraction. The condensate and volatile fractions distilling from the absorbed oil were analyzed by standard methods, and the gas was analyzed on a KhT-2M gas-adsorption chromatograph. The hydrogen sulfide was determined by the ferricyanide method [16], and after it had been absorbed the sum of the organic sulfur compounds was determined by the combustion method [17].

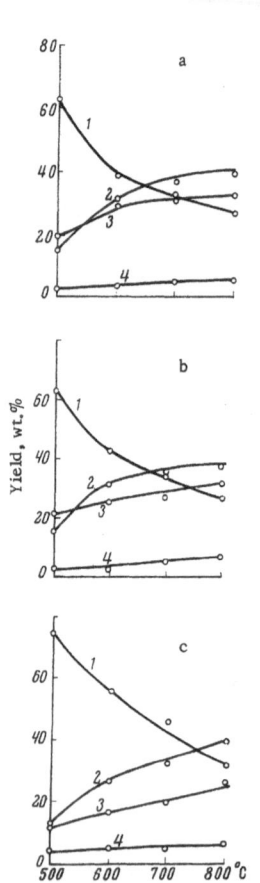

Fig. 2. Yields of products from thermal decomposition of mobile packings of processed petroleum products mixed with soot: PP_c (a), PP_{do} (b), and PP_{vg} (c). 1) liquid product (condensate); 2) gas; 3) solid residue + losses; 4) volatile fraction distilled from absorbed oil.

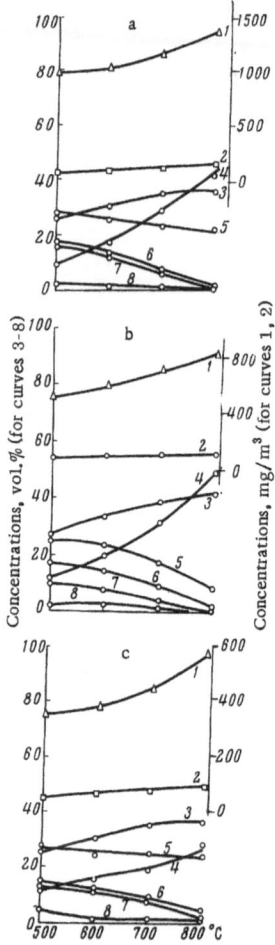

Fig. 3. Compositions of gases from thermal decomposition of PP_c (a), PP_{do} (b), and PP_{vg} (c) on mobile packings. 1) H_2S; 2) organic sulfur compounds (total); 3) CH_4; 4) H_2; 5) C_2H_4; 6) C_3H_6; 7) C_2H_6; 8) C_3H_8.

The combined mass balances of thermal decomposition on the mobile packing are given in Table 3 and plotted in Fig. 2. As we would expect, pyrolysis proceeds further as the reaction-zone temperature increases; as a result, there is a decrease in the yield of liquid products and an increase in the gas, volatile fractions, and solid residue.

TABLE 4. Characteristics of Liquid Products (Condensate) Obtained by Thermal Decomposition of Initial Specimens on Mobile Packings at Various Temperatures

	PP_c				PP_{do}				PP_{vg}			
	Temperature, °C											
	500	600	700	800	500	600	700	800	500	600	700	800
Density ρ_4^{20}, g cm^3..	0.9650	0.9410	0.9200	0.9010	0.9640	0.9650	0.9140	0.8842	0.9775	0.9485	0.9490	0.9070
Refractive index n_D^{20}..	1.5100	1.5338	1.5500	--	1.5005	1.5150	1.5320	1.5450	1.5109	1.5310	1.5390	--
Viscosity (from Engler viscometer)	1.32	1.31	1.22	1.14	1.27	1.12	1.12	1.04	7.40	3.09	2.38	1.83
Iodine number (Kaufmann),.......	74.5	92.4	84.0	101.3	67.5	78.6	84.7	90.2	52.8	63.5	80.0	85.5
Total sulfur, wt.%...	1.41	1.60	1.85	2.04	1.15	1.17	1.20	1.53	0.80	1.02	1.14	1.54
Total sulfonatables, vol. %........	40.0	45.0	57.0	64.0	30.0	37.0	45.0	56.0	42.0	46.0	57.0	61.0
Flash point, °C.....	47.0	40.0	34.0	29.0	92.0	48.0	36.0	37.0	108.0	103.0	92.0	94.0
Ignition point, °C...	61.0	53.0	48.0	38.0	102.0	72.0	55.0	49.0	175.0	149.0	113.0	115.0
Engler distillation, °C												
fbp...........	75	70	68	62	110	107	95	92	110	101	96	90
20% distillate.....	150	118	112	108	238	190	176	163	240	220	208	189
40% distillate.....	220	170	164	154	251	240	234	221	298	273	253	241
60% distillate.....	257	238	230	225	265	264	260	252	--	299	275	271
80% distillate.....	289	263	250	242	290	287	282	276	--	--	305	297
fbp...........	297	279	255	259	301	300	292	280	309	300	305	300

TABLE 5. Characteristics of Volatile Fractions Distilled from Absorbed Oil in Thermal Decomposition of Initial Specimens on Mobile Packing

Initial specimen	Temp,, °C	Density ρ_4^{20},g/cm^3	Refractive index, n_D^{20}	Iodine No (Kaufmann)	Total sulfur, wt.%	Sum of sulfonatables (Katwinkel), vol.%
PP_c	500	0.7550	1.4242	17.9	0.17	4.6
	600	0.7560	1.4255	18.5	0.19	10.4
	700	0.7820	1.4440	48.4	0.21	11.6
	800	0.7860	1.4450	49.1	0.20	12.6
PP_{do}	500	0.7545	1.4220	26.6	0.15	6.0
	600	0.7580	1.4278	37.4	0.19	6.8
	700	0.7652	1.4328	69.0	0.20	8.0
	800	0.7990	1.4485	73.0	0.33	12.0
PP_{vg}	500	0.7400	1.4210	36.0	0.10	4.3
	600	0.7675	1.4400	73.5	0.20	9.0
	700	0.7824	1.4465	76.0	0.24	11.2
	800	0.7860	1.4520	71.6	0.31	14.0

At the same time (see Fig. 3), as the temperature rises we observe an increase in the percentage content of hydrogen and methane and a reduction in the total content of low olefins. The absolute yield of the latter increases, however.

The content of hydrogen sulfide and organic sulfur compounds in the gas increases with temperature, but their total amount is less than we would expect from the results of semicoking in an aluminum retort. This is

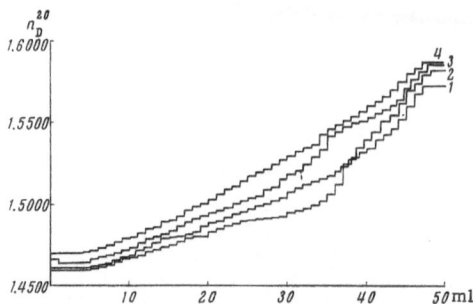

Fig. 4. Chromatograms of liquid products (condensate) obtained
for various temperatures of decomposition of processed vacuum—
gas—oil mixed with soot (PP_{vg}) on mobile packings. Temperature in °C: 1) 500; 2) 600; 3) 700; 4) 800.

explained by the presence of large amounts of cast-iron packing in the reaction zone; in these conditions certain reactions with sulfur-containing compounds may occur on the cast-iron surfaces. This guess is confirmed
by the results of analysis of packing withdrawn from the cycle. In addition, it was found that if petroleum products mixed with cast-iron balls are semicoked in an aluminum retort, there is also a marked reduction in the
amount of sulfur compounds in the gas.

The condensate in the receiver and the volatile fractions distilled from the absorbed oil were analyzed by
standard methods. The results are given in Tables 4 and 5. In all cases the refractive index, iodine number,
total sulfonatables, and total percentage sulfur content increased with the reaction-zone temperature.

For example, by displacement chromatography we analyzed the condensate obtained by thermal decomposition of a mixture of vacuum gas—oil with soot (PP_{vg}). Figure 4 gives the resultant chromatograms, which
clearly illustrate the dynamics of the growth of the refractive index as the reaction-zone temperature rises.

From the mass balances and corresponding analysis data we calculated and plotted the distribution of sulfur in the system versus the reaction-zone temperature (Fig. 5). It will be seen that as the temperature rises
there is an increase in the amount of sulfur going into the gas and solid residue and a reduction in the amount
left in the condensate.

We must emphasize that the product yields, the qualitative product characteristics, and the sulfur distribution are determined not only by the temperature but also, to some extent, by the packing material (in our
case, cast-iron balls); as remarked above, on the surface of the packing there may occur various reactions, including those with sulfur-containing compounds.

Not only metal packing, but also other materials (small coke, various ores, ceramic materials, etc.) can
be used in the process, provided that they possess the required mechanical and thermal strength. One very
promising line is the use of small coke particles obtained in the process itself; most of this will be derived
from the working cycle.

SUMMARY

1. It is possible to use semicoking (low-temperature carbonization) and thermal decomposition on mobile
circulating packing for processing liquid electrocracking products mixed with soot.

2. As the temperature rises from 500 to 800°C, there is an increase in the gas and solid-residue yields
and a decrease in the yield of liquid products (condensate).

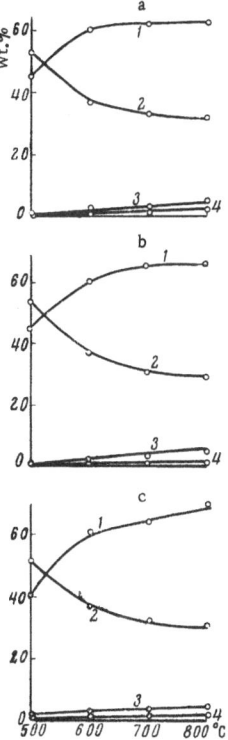

Fig. 5. Distribution of sulfur (in wt.% of total sulfur content) in components obtained by thermal decomposition of processed petroleum products mixed with soot on mobile packing — PP_c (a), PP_{do} (b), and PP_{vg} (c). 1) Solid residue + losses; 2) liquid product (condensate); 3) gas; 4) volatile fractions distilled from absorbed oil.

3. The liquid products are characterized by thermal decomposition and the gases obtained.

4. The authors plot graphs of the sulfur distribution in the system versus the temperatures used.

5. Liquid products, obtained from thermal processing (by semicoking or on mobile packings) can be returned for further electrocracking.

6. The gas, which contains appreciable quantities (up to about 35 vol.%) of low olefins, can be utilized in the chemical industry.

LITERATURE CITED

1. Pechuro, N. S., A. N. Merkur'ev, and G. A. Grishin, Symposium: Synthesis and Properties of Monomers, Izd. Nauka, Moscow (1964), p. 22.
2. Pechuro, N. S., É. Ya. Grodzinskii, and O. Yu. Pesin, Symposium: Problems of Electrical Processing of Materials, No. 4, Izd. Akad. Nauk SSSR, Moscow (1962), p. 192.
3. Pechuro, N. S., É. Ya. Grodzinskii, and O. Yu. Pesin, Ibid., p. 209.
4. Grodzinskii, É. Ya., Author's Abstract of Candidate's Dissertation, MITKhT im. M. V. Lomonosova (1964).
5. Lapitskaya, O. I., and N. I. Alekhina, Symposium: Sulfur-Bearing Oils and Products of Their Processing, Izd. Khimiya, Moscow (1964), p. 54.
6. Igonin, P. G., and I. D. Desyatkova, Coking of Petroleum Residues. Symposium: Processing of Petroleum Residues. Izd. GOSINTI, Moscow (1957),p. 110.
7. Kosterev, V. P., Novosti Neft. Tekhn., No. 1:6 (1957).
8. Frangulyan, A. M., Khim. i Tekhn., Topliv i Masel, No. 3:43 (1958).
9. Rapoport, I. B., Artificial Liquid Fuel. Gostoptekhizdat, Moscow (1955), p. 161.
10. Pechuro, N. S., and Z. É. Lider, Russian Patent No. 10691 (1941).
11. Butkov, N. A., and N. A. Shchekina, Neftyanoe Khoz., No. 8-9:40 (1945).
12. Aliev, V. S., Processing Petroleum Residues. Izd. GOSINTI, Moscow (1957).
13. Amerik, B. K., Proceedings of Second Scientific Conference on Fuel Processing. GrozNII, Groznyi (1956), p. 65.
14. Gurvich, V. L., and E. V. Smidovich, The Catalytic Cracking "Chimney" Abroad, Izd. GOSINTI, Moscow (1960).
15. Karsmit, U. E., Oil Forum, No. 3:62 (1957).
16. Blazhenova, A. N., A. A. Il'inskaya, and F. M. Rapoport, Analysis of Gases in the Chemical Industry, Goskhimizdat, Moscow (1953).
17. Dement'eva, M. I., Analysis of Hydrocarbon Gases. Gostoptekhizdat, Moscow-Leningrad (1953), p. 244.

CHANGES IN THE PHYSICOCHEMICAL PROPERTIES
OF ORGANIC LIQUIDS DUE TO PARTIAL DECOMPOSITION
BY A NONSTATIONARY ELECTRICAL DISCHARGE

A. N. Merkur'ev, I. A. Pesina, and N. S. Pechuro

The decomposition of organic liquids in various types of electrical discharge is interesting as a method of obtaining gases rich in unsaturated compounds — in particular, acetylene. However, most research on this subject has been devoted to finding out the optimum conditions corresponding to maximum acetylene yield with minimum power expenditure. Much less attention has been paid to the changes which occur in the physicochemical properties of organic liquids acted on by electrical discharges [1-3].

In this article we shall give the results of some research on the changes which occur in the physicochemical properties of organic products when they are partially decomposed by low-voltage nonstationary discharges. At this stage of the work, our results are merely qualitative, and do not reveal much about the quantitative changes which occur in these organic products.

Our original media were petroleum products No. 1 (ibp 40°C, fbp 160°C), No. 2 (ibp 150°C, fbp 252°C), No. 3 (ibp 127°C, fbp 256°C), and No. 4 (ibp 145°C, fbp 340°C), together with n-octane, n-nonane, n-decane, benzene, and m-xylene.

These organic liquids were electrocracked in a laboratory plant (which was described in [4]) under a no-load voltage of U_{nl} = 60 V. The intermediate contacts were three graphite cylinders (d = 20 mm, H = 20 mm).

Tables 1 and 2 list the physicochemical properties of the liquid organic products before and after electrocracking. Products Nos. 1, 2, and 3 were obtained after the evolution of 250, 500, and 750 liters of gas, respectively, from 2500 ml of initial organic liquid. On comparing the initial and final properties, we see that electrocracking causes changes in the density, refractive index, iodine number, amount of high-boiling fractions, and viscosity. These characteristics increased with the amount of gas evolved, i.e., with increasing decomposition of the initial material.

The resultant liquid electrocracking products are unstable: they darken when exposed to light, and on brief standing a dark-brown precipitate appears in them.

We also found that the formation of this precipitate is hastened by heating the liquid products to 100°C. From the element composition (C 95.5%, H 4.5%) of the precipitate, which was insoluble in hot benzene, we inferred that it consisted of a mixture of carboids and carbenes.* Furthermore, from extracts in benzene, carbon tetrachloride, and acetone, after removal of the extractant, we separated asphaltenes and neutral resins.

We made a more detailed study of the changes in composition of the liquid organic products by displacement chromatography. The adsorbents were Mark ASM silica gel and A-59 alumina gel. The desorbent was

---

* Carboid = fraction insoluble in carbon disulfide; carbene = fraction soluble in carbon disulfide but insoluble in carbon tetrachloride.

TABLE 1. Characteristics of Liquid Organic Products before and after Electrocracking

	Pet. prod., ibp 40°C, fbp 160°C		Pet. prod., ibp 150°C, fbp 252°C				Pet. prod., ibp 127°C, fbp 256°C			Pet. prod., ibp 145°C, fbp 340°C		
	Initial	№ 1	Initial	№ 1	№ 2	№ 3	Initial	№ 1	№ 2	Initial	№ 1	№ 2
Density ρ_{20}^{20}, g/cm^3	0.7324	0.7361	0.7896	0.7928	0.7951	0.7995	0.8203	0.8263	0.8286	0.8413	0.8425	0.8445
Refractive index, n_D^{20}	1.4126	1.4260	1.4390	1.4416	1.4438	1.4457	1.4583	1.4623	0.4631	1.4669	1.4691	1.4711
Viscosity (Engler)	0.93	0.95	1.04	1.05	1.06	1.10	1.08	1.10	1.14	1.11	1.37	1.42
Iodine number (Kaufmann)	74.06	80.75	5.06	6.85	8.64	10.73	49.37	83.50	96.92	9.80	34.50	43.92
Engler distillation, °C												
ibp	40	55	150	150	151	165	127	130	152	145	190	205
distillate 10%	62	68	165	167	170	170	174	185	202	198	230	238
» 20 »	73	79	170	171	174	176	189	195	205	220	243	248
» 30 »	84	91	177	178	181	182	204	205	213	239	256	259
» 40 »	95	103	183	184	188	190	214	215	218	241	264	271
» 50 »	107	113	188	193	195	198	220	221	223	259	272	282
» 60 »	118	126	199	201	203	206	226	229	230	271	280	290
» 70 »	133	141	207	211	214	215	233	235	238	280	291	297
» 80 »	144	157	216	225	226	240	243	246	246	292	304	321
» 90 »	155	169	240	243	245	248	252	255	261	317	322	344
» 95 »	160	178	252	253	256	260	256	265	272	340	353	376

Note: Products Nos. 1, 2, and 3 were obtained after separation of 250, 500, and 750 liters of gas from the original petroleum product, respectively.

TABLE 2. Characteristics of Individual Hydrocarbons before and after Electrocracking

	n-Octane			n-Nonane					n-Decane		
	Initial			Initial					Initial		
	Literature data	Experimental data	№ 1	Literature data	Experimental data	№ 1	№ 2	№ 3	Literature data	Experimental data	№ 1
Density ρ_{20}^{20}, g/cm^3	0.7030	0.7030	0.7075	0.7180	0.7178	0.7215	0.7232	0.7289	0.7300	0.7301	1.7352
Refractive index, n_D^{20}	1.3974	1.3975	1.3998	1.4036	1.4052	1.4064	1.4073	1.4087	1.4120	1.4120	0.4149
Viscosity (Engler)	--	0.96	0.98	--	0.97	0.98	0.98	1.00	--	1.00	1.03
Iodine number (Kaufmann)	--	--	4.25	--	--	6.53	10.45	17.34	--	--	4.70
Engler distillation, °C:											
ibp	125.7	125.7	125.0	150.8	150.4	150.5	150.5	150.5	174.0	174.2	174.5
distillate 10%	--	--	125.5	--	--	150.8	150.7	150.7	--	--	174.5
» 20 »	--	--	125.6	--	--	150.4	150.6	150.6	--	--	174.5
» 30 »	--	--	125.7	--	--	150.9	150.7	150.8	--	--	175.6
» 40 »	--	--	125.8	--	--	150.9	151.0	151.2	--	--	175.8
» 50 »	--	--	125.9	--	--	150.9	151.0	151.6	--	--	175.8
» 60 »	--	--	126.0	--	--	151.6	151.8	152.0	--	--	176.0
» 70 »	--	--	126.2	--	--	152.0	151.9	152.0	--	--	176.0
» 80 »	--	--	126.6	--	--	152.5	152.7	152.5	--	--	177.0
» 90 »	--	--	128.0	--	--	153.2	153.4	153.5	--	--	178.3
» 95 »	--	--	129.1	--	--	151.2	151.4	154.6	--	--	179.4

Note: Products Nos. 1, 2, and 3 were obtained after separation of 250, 500, and 750 liters of gas from the original petroleum product, respectively.

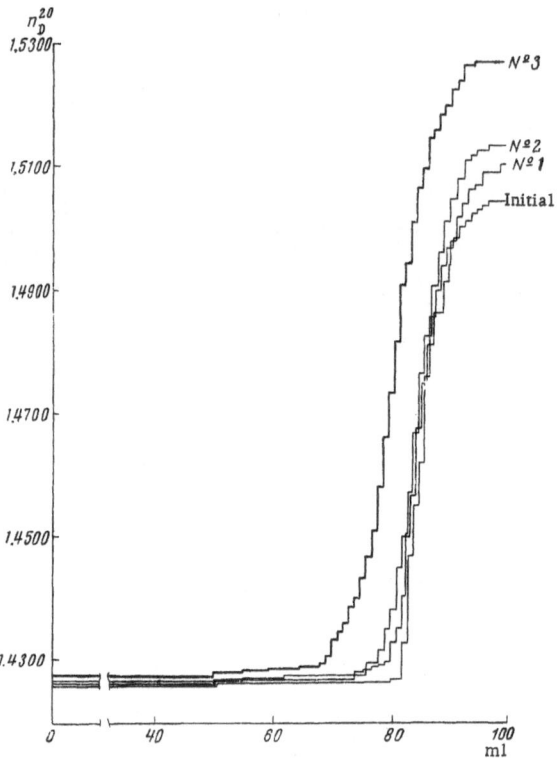

Fig. 1. Chromatogram of initial petroleum product (ibp 150°C, fbp 252°C),
and chromatograms of electrocracking products Nos. 1, 2, and 3.

96% ethyl alcohol. Other desorbents were found to be unsuitable in this case, because, when working with pentane or petroleum ether, we found that considerable amounts of asphaltene-resiny substances were deposited on the adsorbent. Furthermore, as remarked above, even slight heating of the liquid products was accompanied by the formation of a solid precipitate.

We poured 100 ml of each product into an adsorption column. At the outlet it was divided into 100 separate samples of 1 ml each.

These samples were arbitrarily grouped into four fractions according to their refractive indices, aromatics contents, and color: the fractions were designated as "clear" and "yellow" paraffin—naphthenic, transitional, and aromatic. Each of these four fractions was analyzed separately.

Figure 1 shows that the amounts of the fractions with high refractive index increase with the gas yield. Studies of the initial petroleum product (ibp 150°C, fbp 252°C) and the products of its partial decomposition in a nonstationary discharge revealed that the following changes take place as the organic liquid is progressively treated (see Table 3):

1. The content of the "clear" paraffin—naphthenic fraction decreases, while that of the "yellow" and aromatic fractions increases.

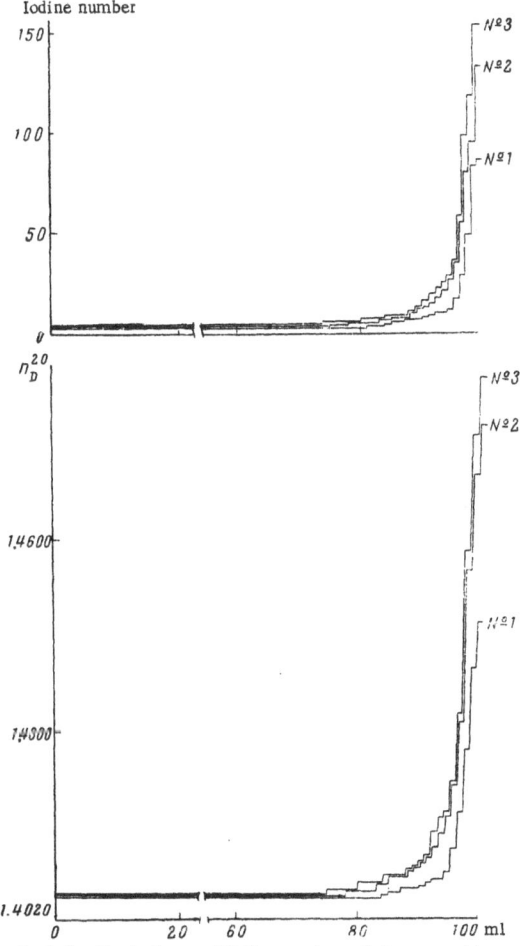

Fig. 2. Refractive indices and iodine numbers of electrocracking products of n-nonane.

2. The density, iodine number, and refractive index of the aromatic fractions all increase.

The "clear" paraffin–naphthene fractions did not contain unsaturated compounds. These did begin to appear in the "yellow" paraffin–naphthene fractions.

Interest attaches to the chromatograms of the liquid organic products obtained by partial decomposition of single compounds. From Fig. 2 we find that the electrocracking of n-nonane gives rise to compounds with higher refractive indices and iodine numbers. Detailed study of the individual fractions revealed that part (2-3%) of the n-nonane forms iso-compounds on decomposition. In these conditions n-nonane also forms naphthenic compounds, which, however, we did not separate into a single fraction.

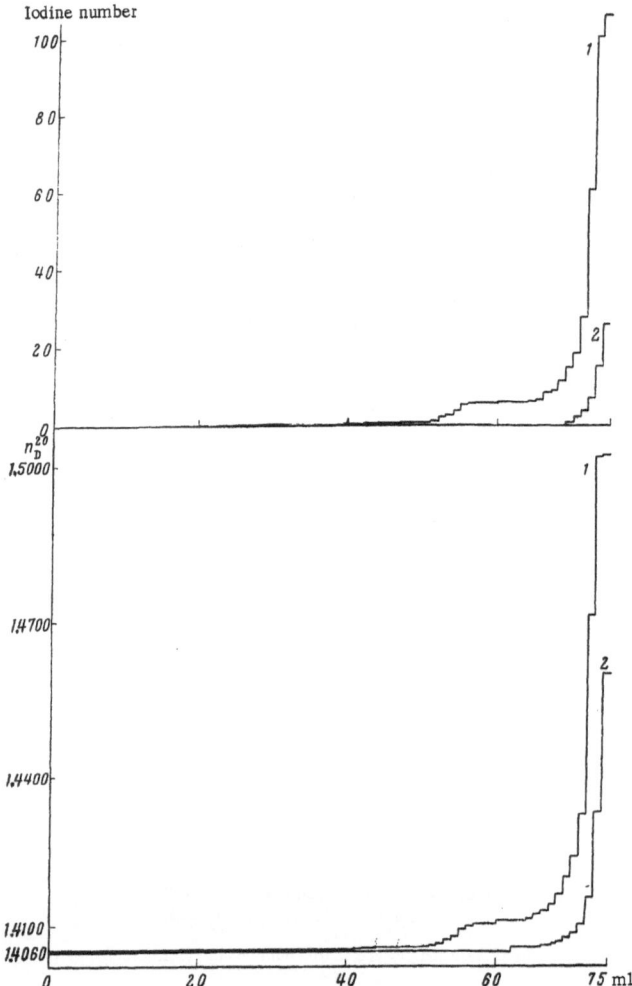

Fig. 3. Refractive indices and iodine numbers of products obtained by electrocracking of n-nonane (1) before and (2) after processing them with mercury acetate.

To elucidate the characters of the unsaturated compounds formed by electrocracking of n-nonane, part of
the product was treated with a solution of mercury acetate. The product remaining after partial separation of
olefins was separated in a chromatographic column. From Fig. 3 we see that the refractive index of the corresponding fractions and their iodine numbers both decrease. We can infer that aromatic compounds with double
bonds in their side chains are formed in the electrocracking products of the paraffins, in particular of n-nonane.

TABLE 3. Characteristics of Separate Fractions of Initial Material (ibp 150°C, fbp 252°C) and of Electrocracking Products Nos. 1, 2, and 3

Index	Fraction	Liquid organic products			
		Initial	No. 1	No. 2	No. 3
Refractive index, n_D^{20}	"Clear" paraffin—naphthenic	1.4257	1.4265	1.4266	1.4270
	"Yellow" paraffin— naphthenic	—	1.4273	1.4283	1.4290
	Transitional	1.4680	1.4580	1.4582	1.4580
	Aromatic	1.5020	1.5090	1.5094	1.5098
Density, ρ_{20}^{20}, g/cm³	"Clear" paraffin—naphthenic	0.7692	0.7696	0.7699	0.7705
	"Yellow" paraffin—naphthenic	—	0.7705	0.7714	0.7717
	Transitional	0.8133	0.8190	0.8134	0.8180
	Aromatic	0.8847	0.8870	0.8905	0.8944
Iodine number (Kaufmann)	"Clear" paraffin—naphthenic	0.20	0.14	0.09	0.11
	"Yellow" paraffin—naphthenic	—	1.33	1.42	2.29
	Transitional	15.80	15.93	14.87	15.88
	Aromatic	21.60	22.79	24.67	25.08
Contents of fractions, vol.%	"Clear" paraffin—naphthenic	83.0	75.0	72.0	64.0
	"Yellow" paraffin—naphthenic	—	5.0	4.5	8.0
	Transitional	8.0	5.0	7.0	7.0
	Aromatic	12.0	15.0	16.5	21.0

Note: Products Nos. 1, 2, and 3 were obtained after separation of 250, 500, and 700 liters of gas, respectively, from the initial petroleum product.

From an analysis of the products obtained by partial decomposition of aromatic compounds (benzene, m-xylene), we can infer that in this case there is an increase in the refractive index of the individual fractions, accompanied by a rise in iodine numbers. No paraffin or naphthene compounds were observed in the electrocracking products of benzene and m-xylene.

In conclusion we should remark that, owing to the complexity and instability of the electrocracking products of the liquid organic feedstock, we were unable to separate them sharply. However, we found that all the organic liquids which we examined, when acted on by nonstationary low-voltage electrical discharges, form some aromatic and olefin compounds, together with condensation products of the asphaltene type. We also established experimentally that similar changes take place in liquid organic products subjected to arc discharges.

SUMMARY

1. When an electrical discharge acts on a liquid organic medium, aromatic and unsaturated compounds appear in the product, together with condensation products of the asphaltene, carboid, and carbene types.

2. The yields of these products increase with the extent of decomposition, assessed by the amount of gas evolved.

LITERATURE CITED

1. Kokurin, A. D., O. E. Setkina, and V. V. Gruzdeva, Tr. Leningrad. Tekhn. Inst. im. Lensoveta, 57:113 (1959).
2. Andrussow, L., Erdöl u. Kohle, 12:24 (1959).
3. Kroepelin, H., Chem. Ind. Techn., 28:703 (1956).
4. Pechuro, N. S., and A. N. Merkur'ev, Symposium: Problems in the Electrical Treatment of Materials, No. 4, Izd. Akad. Nauk SSSR, Moscow (1962), p. 181.

THE INFLUENCE OF CURRENT STRENGTH AND PULSE LENGTH
ON THE DECOMPOSITION OF AN ORGANIC LIQUID
BY A CONDENSED DISCHARGE

O. Yu. Pesin, A. N. Merkur'ev, and N. S. Pechuro

When an electrical discharge takes place in a liquid dielectric, as well as decomposition of the material, evolution of gas, and formation of solid condensation products (soot), we also observe some erosion of the electrodes. Electrical discharges, in addition to their use for obtaining gases rich in unsaturated compounds such as acetylene from various types of liquid hydrocarbon feedstock [1-3], are thus also used for electroerosion (spark machining) [4].

It is known [5, 6] that to get maximum yields of cracking gas and acetylene, the energy in the discharge gap must be supplied in definite pulses. This facilitates evacuation of the gaseous products from the discharge zone and reduces the probability of their further decomposition by subsequent pulses. Obviously, the gas yield will increase with the energy supplied to the discharge gap. However, since the discharge energy depends on the working voltage, the current strength, and the pulse length, it is of theoretical and practical interest to find out which of these parameters is crucial for the gas yield. In a previous paper [7], we discussed the influence of the working voltage U_p and current strength I_p on the electrocracking process, and showed that, for a high-voltage ac arc with a power in the range from about 7 to 56 W, the gas yield is mainly determined by the discharge power (i.e., $U_p \cdot I_p$), and it does not matter whether we use a low current and a high voltage, or vice versa.

TABLE 1. Pulse Parameters Obtained with Various Inductances in Discharge Circuit

Pulse parameter	Number of turns of coil, n				
	0	5	10	15	20
Pulse energy W, j	0.207	0.20	0.190	0.184	0.188
	0.830	0.80	0.805	0.850	0.806
	1.430	1.37	1.390	1.450	1.400
Pulse duration τ, μsec. . .	67.6	102.7	150.0	207.7	269.2
	161.5	215.4	299.9	392.2	484.6
	200.0	260.0	390.0	461.5	576.9
Current strength I, A . . .	185.0	122.0	93.6	73.5	57.8
	292.6	232.7	171.2	147.1	126.3
	361.3	320.0	230.0	200.0	170.2

Note: In all cases the working voltage was practically constant and equal to 32-35 V.

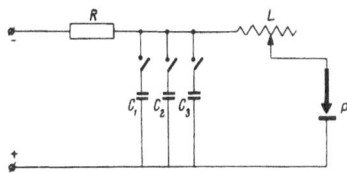

Fig. 1. Schematic circuit of apparatus.

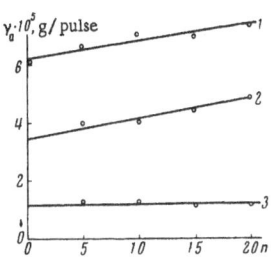

Fig. 2. Erosion of anode γ_a versus number of turns n on inductance coil for various pulse energies W. W (in joules): 1) 1.4; 2) 0.8; 3) 0.2.

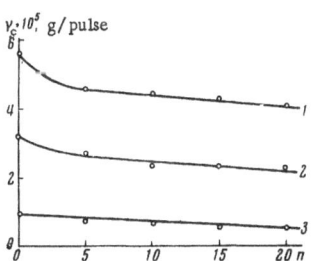

Fig. 3. Erosion of cathode γ_c versus number of turns n on inductance coil, for various pulse energies. Notation: see Fig. 2.

In this article we shall see how the pulse length τ and current strength I affect the decomposition of n-undecane in a condensed discharge. We used a laboratory plant with a gastight bath, a steel anode, and a brass cathode. The gas formed was measured with a GSB-400 meter, and was analyzed with a KhT-2M chromathermograph. The discharge parameters were measured by means of an oscillograph.

The schematic circuit of the apparatus is shown in Fig. 1. It comprises a constant-current source, a set of capacitances C_1-C_3, a discharge resistance R, and an inductance coil L. It permits the formation of pulses in the discharge gap P which have roughly equal energies, but which differ in current strength and duration. We used three fixed energies, W = 0.2, 0.8, and 1.4 J. The pulse parameters were varied by appropriate selection of the capacitance and by varying the number of turns on L from 5 to 20, thus varying the inductance from 10 to 55 μH. The pulse parameters are listed in Table 1.

The reader will see that an increase in pulse energy leads to a rise in duration and current strength, independent of the inductance in the discharge circuit. For constant pulse energy, an increase in inductance causes a drop in current and an increase in pulse length. We thus studied pulses ranging in length from 67.6 to 576.9 μsec and in current from 57.8 to 361.3 A. Since pulses with these parameters are used in electric spark machining, it was of interest to elucidate the principal laws governing electrode wear in these conditions, in addition to our studies of the mutual influence of current strength and pulse length on the yield of gaseous decomposition products. Our results are plotted in Figs. 2-5.

We easily see from these graphs that, for any pulse energy, an increase in the inductance of the discharge circuit is accompanied by some increase in anode wear and decrease in cathode wear; however, the total erosion remains practically constant.

The relative cathode wear γ_c / γ_a decreases quite markedly as the inductance increases.

Figures 6-9 give data characterizing the process of decomposition of the original product. They show that increase in the inductance of the discharge circuit (W = const) leads to reduction in the yields of gaseous and solid decomposition products (soot), and also to reduction in the amount of n-undecane decomposed per pulse. In addition, there is a decrease in the proportion of the pulse energy which is expended on the chemical process η; this was calculated from the overall equation of the decomposition reaction and the thermal effect of this reaction Q_p. This latter was nearly constant and equal to about 190 kcal/mole. The composition of the product gas was also nearly constant: 18-20 vol.% C_2H_2, 3-6 vol.% CH_4, 14-16 vol.% C_2H_4, and 58-62 vol.% H_2.

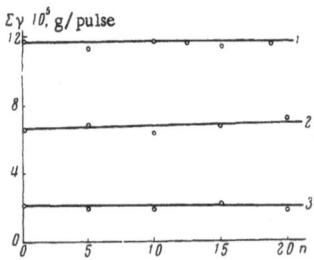

Fig. 4. Total erosion of electrodes $\Sigma\gamma$ versus number of turns n on inductance coil, for various pulse energies. Notation: see Fig. 2.

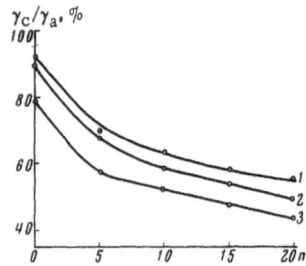

Fig. 5. Relative cathode wear γ_c/γ_a versus number of turns n on inductance coil, for various pulse energies. Notation: see Fig. 2.

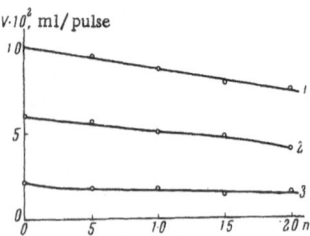

Fig. 6. Gas yield V versus number of turns on inductance coil, n, for various pulse energies. Notation: see Fig. 2.

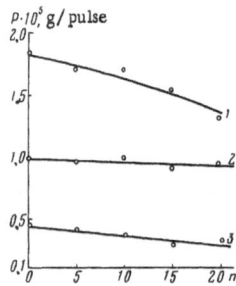

Fig. 7. Soot yield P versus number of turns n on inductance coil, for various pulse energies. Notation: see Fig. 2.

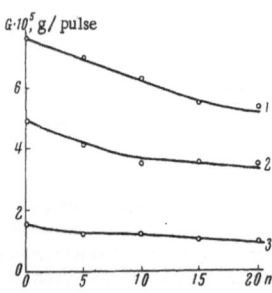

Fig. 8. Quantity of feedstock decomposed G, versus number of turns n on inductance coil, for various pulse energies. Notation: see Fig. 2.

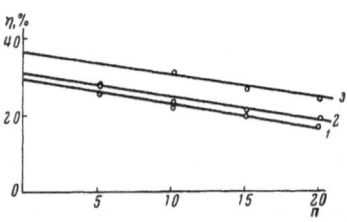

Fig. 9. Coefficient η versus number of turns n on inductance coil, for various pulse energies. Notation: see Fig. 2.

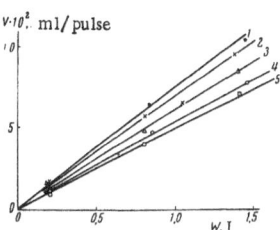

Fig. 10. Gas yield V versus pulse energy W and number of turns on inductance coil n. Values of n: 1) 0.2; 2) 5; 3) 10; 4) 15; 5) 20.

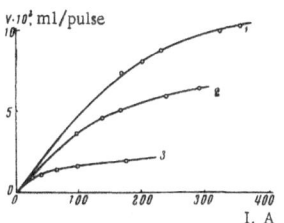

Fig. 11. Gas yield V versus current strength for various pulse energies. Notation: see Fig. 2.

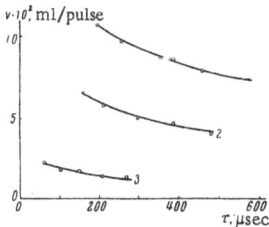

Fig. 12. Gas yield V versus pulse length τ for various pulse energies. Notation: see Fig. 2.

Since the soot yield, raw-material consumption, and coefficient η are closely linked to the amount of gas evolved, it was of interest to see how this quantity varies with the pulse energy for various discharge-circuit inductances (Fig. 10). We found that for each series of experiments (n = const) the gas yield is directly proportional to the pulse energy. Since the slope of the line plotting V versus W decreases as the inductance increases, the way in which we achieve a given energy is far from unimportant: it makes a difference whether we use a high current and low duration, or vice versa. However, our results do not so far unambiguously show which factor predominates in determining the gas yield.

Figures 11 and 12 plot the gas yield versus the current strength and pulse length with W = const. They show that an increase in current strength leads to a rise in gas yield, while, on the other hand, an increase in pulse length reduces the gas yield. However, it must be remembered that, in our conditions, an increase in I (with W = const) automatically leads to a reduction in τ, and a rise in τ is accompanied by a fall in I; in this case, therefore, we have elucidated only the total effect of both these factors together.

To decide which of these two parameters, I or τ, is crucial in governing the formation of large amounts of gas, we can proceed by comparing the gas yields when the same amount of energy is expended in two ways — with increased current strength at constant pulse length, or with increased pulse length at constant current strength. The optimum variant will be the one which gives the greater amount of gas for unit expenditure of energy.

The comparison can be made from our results. Figure 13 plots the gas yield versus the pulse energy for several constant values of I and τ, lying within the ranges which we studied. These graphs were constructed on the basis of the data in Figs. 11 and 12. From Fig. 13 we see that as the pulse energy increases, either as a result of increased current strength (τ = const) or increasing pulse length (I = const), the gas yield rises.

Thus it is now clear why the graph of gas yield versus current (W = const) displays rising curves tending to a constant limit (see Fig. 11). This is clearly due to the fact that, when W = const, a rise in current (accompanied by an increase in gas yield) leads to a simultaneous decrease in the pulse length, which in turn (as shown by Fig. 13) causes a reduction in the yield of gaseous products. Obviously, for a certain current strength, determinate for each value of W, these two phenomena will mutually compensate out, and the gas yield will be almost constant.

In addition, Fig. 13 does not only demonstrate that the gas yield increases with I and τ, but also shows which of these parameters will have the greater effect.

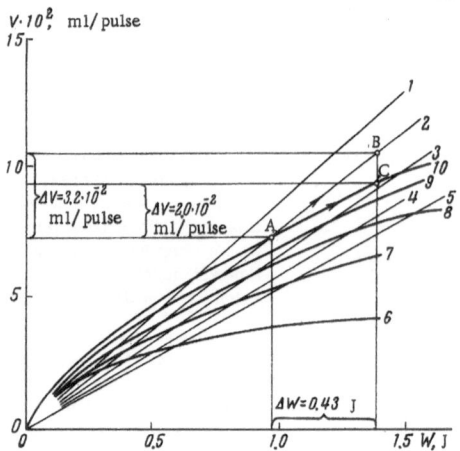

Fig. 13. Gas yield V versus pulse energy W for various currents I and pulse lengths τ. τ (in μsec): 1) 120; 2) 200; 3) 300; 4) 400; 5) 500. I (in A): 6) 100; 7) 150; 8) 200; 9) 250; 10) 300.

Let us consider point A, which has the following values: W = 0.97 J, I = 300 A, τ = 200 μsec, V = $7.4 \cdot 10^{-2}$ ml pulse. We see from the graph that, if we increase the pulse energy, for example, to 1.4 J (point B), with the same value of τ (i.e., by increasing the current strength), then we get a rise in gas yield, $\Delta V = (10.60 - 7.40) \cdot 10^{-2} = 3.20 \cdot 10^{-2}$ ml/pulse, or $3.20 \cdot 10^{-2} : (1.40 - 0.97) = 7.44 \cdot 10^{-2}$ ml/J. At constant current strength, an increase of energy (point C) gives: $\Delta V = (9.40 - 7.40) \cdot 10^{-2} = 2.00 \cdot 10^{-2}$ ml/pulse, or $2.00 \cdot 10^{-2} : (1.40 - 0.97) = 4.55 \cdot 10^{-2}$ ml/J.

We easily see that this phenomenon is due to the fact that the (V, W) curves at constant τ are steeper than the (V, W) curves at constant I. The rule must therefore be the same, not only at point A, but also at any other point — as we quickly see from Fig. 13.

Thus, for the same expenditure of energy, a greater gas yield will be obtained by keeping τ constant and increasing the current, rather than by keeping I constant and increasing the pulse length.

SUMMARY

1. We studied gas yield in relation to current strength and pulse length in the decomposition of n-undecane in a condensed discharge.

2. Increases in either pulse length (at constant current) or current strength (at constant pulse length) lead to increased gas yield; however, the current strength is the more decisive factor.

3. The reduction in gas yield accompanying rise in discharge-circuit inductance is due to the reduction in current strength rather than to the rise in pulse length.

LITERATURE CITED

1.   Contardi, A., Giorn. Chim. Ind., No. 7:195 (1925).
2.   Tatarinov, V. V., Russian Patent No. 39904 (1934).
3.   Kroepelin, H., Chem. Ind. Techn., 28:2419 (1956).

4. Lazarenko, B. R., and N. I. Lazarenko, Spark Machining of Conducting Materials. Izd. Akad. Nauk SSSR, Moscow (1958), p. 184.
5   Roginskii, S. Z., and A. B. Shekhter, Zh. Prikl. Khim., 10:473 (1937).
6. Pechuro, N. S., and A. N. Merkur'ev, Symposium: Problems in the Electrical Treatment of Materials, No. 4, Izd. Akad. Nauk SSSR, Moscow (1962), p. 181.
7. Grodzinskii, É. Ya., N. S. Pechuro, and O. Yu. Pesin, Symposium: Spark Machining of Metals, Izd. Akad. Nauk SSSR, Moscow (1963), p. 100.

DECOMPOSITION OF SINGLE HYDROCARBONS DURING
ELECTRO EROSION POWERED
BY A MECHANICAL PULSE GENERATOR

N. S. Pechuro, V. I. Gol'din, and A. N. Merkur'ev

The twenty-year history of electroerosion has seen many publications on the physical theory of the process, the development of power sources, the advances in equipment design, the process technology, the decontamination of the dielectric, etc. However, there has been hardly any study of the way in which the physicochemical properties of the interelectrode dielectric medium affect the process parameters. The organic media used are mainly various petroleum products (kerosene, oils, etc.), which are, of course, complex mixtures of paraffins, naphthenes, and aromatic hydrocarbons. The individual properties of the components would therefore be expected to have different effects on their decomposition and on the technological characteristics of the process. In some cases this has been confirmed experimentally.

Pechuro et al. [1] showed that during spark machining of metals there is extensive decomposition of the medium, producing gas and condensation products (soot). The gases contain up to 20.0 vol.% of acetylene (as compared with 30.0-35.0% in electrocracking). Physicochemical changes occur in the dielectric: the density of the petroleum product, synthine, or shale tar increases, together with their flash point, viscosity, refractive index, and content of sulfonatables. All this is from just one investigation of the physicochemical changes in liquid dielectrics during electroerosion. We must also mention the work of some Polish investigators [2], who used individual organic compounds as the interelectrode medium; however, these workers were interested mainly in the technological characteristics of the spark-machining process.

Thus, further work is needed to study how various classes of organic compounds, when used as interelectrode media, influence the chemical reaction occurring during spark machining.

In our investigation we used the following hydrocarbons as interelectrode media: n-heptane, n-octane, n-decane, tetradecane, cyclohexane, benzene, o-xylene and tetralin. The experimental conditions were close to those prevailing during the machining of holes in steel (St-3) components by a cylindrical brass (L-59) electrode-instrument. The power source was an MIG-2B mechanical pulse generator giving a pulse repetition frequency of $f = 600$ Hz. To get unipolar pulses in the feed circuit, the device included a VK-200 power diode which cut out the reverse voltage half-wave (V_{nl}). Typical current and potential oscillograms are shown in Fig. 1. The details of the experimental method were described in [3].

The oscillograms were processed by the usual method — determination of the main parameters of the operational pulses, averaged over 100 separate measurements: these parameters were the operational voltage V_{av}, pulse length τ_{av} (μsec), mean current I_{av} (A), and pulse energy W_{av} (J).

Depending on the conditions of operation of the power source (which are determined by the current in the generator excitation windings I_{exc}, A), the mean pulse length was varied between 700 and 1100 μsec, and the operative voltage between 20 and 30 V; neither of these parameters depends on the nature of the medium.

114

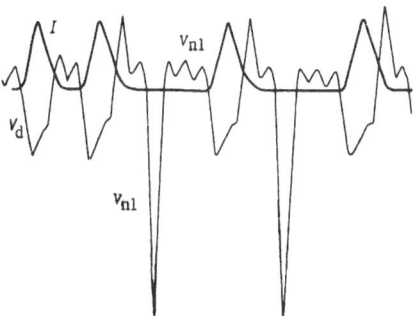

Fig. 1. Current and potential oscillograms of electroerosion process.
I) Current pulse; V_d) voltage pulse during discharge; V_{nl}) no-load pulses
for forward and reverse voltage halfwave. (Dielectric, o-xylene; I_{exc} = 2.0 A.)

TABLE 1. Variation of Energy of Operative Pulses (J) in Relation to Medium Used

I_{exc}, A	n-Heptane	n-Octane	n-De-cane	n-Tetra-decane	Cyclo-hexane	Ben-zene	o-Xy-lene	Tetralin
1.0	1.420	0.800	0.870	0.866	1.23	1.26	1.22	1.26
1.5	2.210	1.930	2.050	1.720	—	2.22	2.16	1.84
2.0	2.290	2.790	2.840	2.060	2.66	2.68	2.53	2.58

TABLE 2. Characteristics of Decomposition of Single Hydrocarbons in Electroerosion

Characteristic	I_{exc}, A	n-Heptane	n-Octane	n-Decane	n-Tetra-decane	Cyclo-hexane	Benzene	o-Xylene	Tetralin
(Absolute gas yield) × 10^2, ml/pulse	1.0	5.59	5.37	6.32	6.04	3.28	2.05	3.24	3.98
	2.0	7.56	11.93	10.90	11.16	4.71	5.09	5.42	7.70
(Relative gas yield) × 10^2, ml/J......	1.0	3.94	6.71	7.26	6.97	2.66	1.62	2.65	3.15
	2.0	3.30	4.27	3.84	5.41	1.77	1.89	2.54	2.98
(Absolute soot yield) × 10^5, g/pulse.....	1.0	1.130	1.040	1.300	0.999	1.060	2.810	2.160	2.480
	2.0	1.670	2.460	1.943	2.250	1.520	4.160	3.970	4.310
(Relative soot yield) × 10^5, g/J........	1.0	0.796	1.300	1.494	1.153	0.879	1.44	1.78	2.00
	2.0	0.729	0.881	0.685	1.092	0.579	1.56	1.57	1.68
(Absolute decomp.) × 10^5, g/pulse.....	1.0	3.790	3.710	4.410	4.440	2.160	2.390	3.230	4.170
	2.0	5.120	8.340	7.760	7.960	3.160	5.950	6.420	8.230
(Relative decomp.) × 10^5, g/J........	1.0	2.669	4.637	5.068	5.080	1.82	1.92	2.61	3.42
	2.0	2.235	2.989	2.732	3.864	1.21	2.26	2.68	3.24

The same also applies to the mean current strength, and hence also to W_{av} (Table 1).

The most probable reason for the scatter in the pulse energies for the various hydrocarbons in corresponding conditions is the conditions of the experiment. The stability of operation of a powerful mechanical pulse-generator plant is governed by the optimality of the conditions (primarily the area under treatment and the boiling point of the dielectric).

　　　　　　　　N. S. PECHURO, V. I. GOL'DIN, AND A. N. MERKUR'EV

TABLE 3. Characteristics of Gases Obtained by Decomposition of Single Hydrocarbons during Electroerosion

Compound	I_{exc}, A	Contents of components, vol.%						
		H_2	CH_4	C_2H_6	C_2H_4	C_3H_8	C_3H_6	C_2H_2
Cyclohexane	1.0	75.00	1.41	--	6.09	--	--	17.50
	2.0	75.30	1.60	--	6.50	--	--	16.60
n-Heptane	1.0	64.32	4.20	0.57	13.45	0.20	1.05	16.21
	1.5	65.00	4.04	0.48	14.20	0.21	0.87	15.20
	2.0	65.65	4.95	0.51	12.90	0.17	1.15	14.67
n-Octane	1.0	63.94	3.48	0.49	13.38	0.11	0.90	17.70
	1.5	62.70	3.65	0.57	14.42	0.11	1.15	17.40
	2.0	63.00	4.04	0.53	15.00	0.28	1.13	16.02
n-Decane	1.0	63.02	3.35	0.40	15.00	0.18	0.80	17.25
	2.0	58.70	4.07	0.54	16.80	0.23	1.35	18.31
n-Tetra-decane	1.0	57.50	3.00	0.42	17.30	--	0.92	20.86
	1.5	60.38	3.17	0.51	15.90	0.13	1.31	18.60
	2.0	61.01	3.44	0.45	14.80	0.11	0.92	19.27
Benzene	1.0	81.70	--	--	0.67	--	--	17.63
	1.5	79.80	--	--	0.80	--	--	19.40
	2.0	75.40	--	--	0.90	--	--	23.70
o-Xylene	1.0	79.60	1.57	--	0.93	--	--	17.90
	1.5	73.30	2.20	--	1.40	--	--	23.10
	2.0	71.20	4.07	--	1.73	--	--	23.00
Tetralin	1.0	68.74	1.48	--	6.28	--	--	23.50
	2.0	61.00	1.50	--	7.30	--	--	30.20

Since we were not using the optimum area, the voltage pulses from the generator were not fully utilized, and this is the cause of the scatter in the parameters.

As will be seen from Table 2, all the parameters characterizing the decomposition of the hydrocarbons under test depend markedly on the class of compounds. The absolute gas yield per pulse, which increased with W_{av} for all the compounds, was half as great for the cyclohexane and aromatic hydrocarbons as for the paraffins. However, the absolute yields of solid products (soot) displayed the opposite relations: the highest results were obtained for the aromatic compounds, those of the paraffins and cyclohexane being much lower.

The data on the yields of solid and gaseous decomposition products show that the hydrocarbons decompose by different routes, and this conclusion is supported by an analysis of the gases. The results of this analysis, shown in Table 3, show that there is little variation in the gas compositions obtained by the decomposition of the paraffins. The hydrogen content varies from 58 to 65 vol.%, the acetylene content from 14 to 20%; there is practically no variation in the concentrations of the remaining components (methane and its homologs and ethylene). When, however, we go over to cyclic and aromatic hydrocarbons, the gas composition shows a marked change. Cyclohexane and tetralin give gases which consist mainly (more than 90%) of hydrogen and acetylene, with very small amounts of saturated compounds and low olefins.

The gas obtained from benzene consists almost entirely of hydrogen and acetylene (the content of low olefins is less than 1%). The gas from o-xylene contains methane in concentrations comparable with those obtained from the paraffins.

The thermal effects of the decomposition reactions, and also the soot yields and quantity of medium decomposed, were calculated from empirical equations based on the gas analysis data; the method of compilation was explained in [1] (see Tables 4 and 5). The thermal effect Q (kcal) was converted to a single pulse q (J/pulse) from the experimental data on the gas yield per pulse $V \cdot 10^2$ (ml/pulse) and the hydrogen concentration in the gas mixture a (vol.%), using the stoichiometric coefficient for hydrogen n_1 in the decomposition

TABLE 4. Results of Thermodynamic Calculations on the Decomposition of Paraffins and Cyclohexane

Medium	I_{exc}, A	Equation of reaction	$-Q$, kcal/mole	$-q$, J/pulse	q/W_{av}, %
n-Heptane	1.0	$C_7H_{16} = 4.229\ H_2 + 0.277\ CH_4 + 0.038\ C_2H_6 + 0.887\ C_2H_4 + 0.013\ C_3H_8 + 0.069\ C_3H_6 + 1.070\ C_2H_2 + 2.487\ C$	116.986	0.1854	13.06
	2.0	$C_7H_{16} = 4.318\ H_2 + 0.325\ CH_4 + 0.033\ C_2H_6 + 0.850\ C_2H_4 + 0.011\ C_3H_8 + 0.075\ C_3H_6 + 0.964\ C_2H_2 + 2.723\ C$	110.100	0.2686	12.15
n-Octane	1.0	$C_8H_{18} = 4.699\ H_2 + 0.262\ CH_4 + 0.037\ C_2H_6 + 1.050\ C_2H_4 + 0.008\ C_3H_8 + 0.068\ C_3H_6 + 1.330\ C_2H_2 + 2.676\ C$	139.627	0.1902	23.78
	2.0	$C_8H_{18} = 4.581\ H_2 + 0.296\ CH_4 + 0.039\ C_2H_6 + 1.098\ C_2H_4 + 0.020\ C_3H_8 + 0.083\ C_3H_6 + 1.185\ C_2H_2 + 2.806\ C$	131.500	0.4023	14.42
n-Decane	1.0	$C_{10}H_{22} = 5.716\ H_2 + 0.304\ CH_4 + 0.036\ C_2H_6 + 1.360\ C_2H_4 + 0.016\ C_3H_8 + 0.073\ C_3H_6 + 1.565\ C_2H_2 + 3.507\ C$	167.341	0.2177	25.03
	2.0	$C_{10}H_{22} = 5.220\ H_2 + 0.357\ CH_4 + 0.047\ C_2H_6 + 1.475\ C_2H_4 + 0.020\ C_3H_8 + 0.119\ C_3H_6 + 1.610\ C_2H_2 + 2.962\ C$	170.372	0.3994	13.71
n-Tetradecane	1.0	$C_{14}H_{30} = 6.957\ H_2 + 0.381\ CH_4 + 0.053\ C_2H_6 + 2.200\ C_2H_4 + 0.024\ C_3H_6 + 2.650\ C_2H_2 + 3.741\ C$	259.670	0.2417	27.92
	2.0	$C_{14}H_{30} = 7.547\ H_2 + 0.427\ CH_4 + 0.056\ C_2H_6 + 1.828\ C_2H_4 + 0.014\ C_3H_8 + 0.113\ C_3H_6 + 2.380\ C_2H_2 + 4.664\ C$	239.600	0.4031	19.57
Cyclohexane	1.0	$C_6H_{12} = 4.254\ H_2 + 0.083\ CH_4 + 0.340\ C_2H_4 + 0.900\ C_2H_2 + 3.437\ C$	88.873	0.0958	7.79
	2.0	$C_6H_{12} = 4.180\ H_2 + 0.089\ CH_4 + 0.361\ C_2H_4 + 0.920\ C_2H_2 + 3.349\ C$	90.111	0.1425	5.52

TABLE 5. Results of Thermodynamic Calculations on the Decomposition of Aromatic Compounds

Medium	I_{exc}, A	Equation of reaction	$-Q$, kcal/mole	$-q$, J/pulse	q/W_{av}, %
Benzene	1.0	$C_6H_6 = 2.434\ H_2 + 0.020\ C_2H_4 + 0.526\ C_2H_2 + 4.908\ C$	17.036	0.0218	1.73
	1.5	$C_6H_6 = 2.373\ H_2 + 0.024\ C_2H_4 + 0.579\ C_2H_2 + 4.794\ C$	19.957	0.0436	1.96
	2.0	$C_6H_6 = 2.242\ H_2 + 0.027\ C_2H_4 + 0.704\ C_2H_2 + 4.538\ C$	26.769	0.0854	3.18
o-Xylene	1.0	$C_6H_4(CH_3)_2 = 3.890\ H_2 + 0.076\ CH_4 + 0.045\ C_2H_4 + 0.868\ C_2H_2 + 6.098\ C$	53.370	0.0686	5.62
	1.5	$C_6H_4(CH_3)_2 = 3.522\ H_2 + 0.106\ CH_4 + 0.067\ C_2H_4 + 1.132\ C_2H_2 + 5.496\ C$	67.903	0.1394	6.45
	2.0	$C_6H_4\ (CH_3)_2 = 3.364\ H_2 + 0.192\ CH_4 + 0.082\ C_2H_4 + 1.088\ C_2H_2 + 5.468\ C$	65.549	0.1661	6.52
Tetralin	1.0	$C_{10}H_{12} = 3.796\ H_2 + 0.081\ CH_4 + 0.351\ C_2H_4 + 1.340\ C_2H_2 + 6.537\ C$	93.941	0.1262	10.02
	2.0	$C_{10}H_{12} = 3.360\ H_2 + 0.083\ CH_4 + 0.402\ C_2H_4 + 1.670\ C_2H_2 + 5.773\ C$	112.435	0.2931	11.36

equation. This method was chosen because hydrogen has the greatest concentration in the gas mixture; the effects of relative errors in the gas analysis are thus reduced to a minimum. The formula adopted is

$$q = 1.865 \cdot 10^{-5} \frac{QaV}{n_1}.$$

In all cases, thermodynamic calculations from the empirical equations thus found for the decomposition show that the process is endothermic. Thus, the decomposition of hydrocarbons in electroerosion is accompanied by the expenditure of part of the energy on the chemical processes ($\eta = q/W_{av}$). The value of η depends largely on the class of organic compounds to which the medium belongs.

TABLE 6. Indices Characterizing the Erosion of the Electrode Material when Single Hydrocarbons are Used as the Interelectrode Medium

Index	$I_{exc.}$ A	n-Heptane	n-Octane	n-Decane	n-Tetradecane	Cyclohexane	Benzene	o-Xylene	Tetralin
(Absolute total erosion of cathode and anode) $\cdot 10^5$, g/pulse.	1.0	10.70	11.03	13.08	12.05	9.15	14.67	16.60	19.04
	1.5	14.48	22.44	20.22	20.64	—	23.81	25.37	30.11
	2.0	16.54	27.45	25.13	24.12	15.22	34.35	34.90	34.10
(Absolute erosion of anode) $\cdot 10^5$, g/pulse	1.0	7.55	7.79	8.88	8.37	6.64	8.40	10.47	11.35
	1.5	9.74	14.87	13.60	13.48	—	13.22	15.35	17.45
	2.0	11.30	18.73	17.85	15.80	11.35	19.25	19.40	19.95
Relative wear of electrode-instrument, wt.%	1.0	41.70	41.59	47.29	43.96	37.80	74.64	58.54	67.75
	1.5	48.90	50.90	48.67	43.11	—	80.02	65.27	72.55
	2.0	46.40	46.55	40.78	52.65	34.09	78.44	79.89	70.92
(Specific total erosion of anode and cathode) $\cdot 10^5$, g/J	1.0	7.54	13.79	15.03	13.91	7.44	11.64	13.60	15.11
	1.5	6.31	11.63	9.86	2.00	—	10.72	11.74	16.36
	2.0	7.49	9.83	8.84	1.71	5.72	12.13	13.79	13.21
(Specific erosion of anode) $\cdot 10^5$, g/J	1.0	5.31	9.73	10.21	19.66	5.40	6.66	8.58	9.01
	1.5	4.25	7.70	6.63	7.84	—	5.95	7.10	9.48
	2.0	5.11	6.71	6.28	7.67	4.26	7.18	7.66	7.73

In the overall balance for the paraffins and cyclohexane (see Table 4), the energy expended on the chemical process varies between 5.52 and 27.92%. The value of η decreases markedly with increasing pulse energy, i.e., in more drastic conditions.

When we come to the aromatic hydrocarbons (see Table 5), the above indices alter considerably. For example, the energy expended on chemical processes is much less for benzene (in mild and drastic conditions) than for cyclohexane. For o-xylene the maximum value of η is somewhat lower than for the paraffins. With tetralin we find an increased expenditure of energy on chemical processes (10.02-11.36%). Thus, for pulse parameters in the ranges which we studied, η varies over a narrower interval for the aromatic compounds than for the paraffins. When we go from mild to more drastic conditions, in the case of the aromatic compounds the proportion of energy expended on chemical processes does not decrease as it does for the paraffins, but rises.

Thus, the physicochemical properties of the hydrocarbons exert some influence on the technological parameters of the process, primarily on the absolute and relative erosion of the electrode materials, and also on the wear ratio (relative electrode wear). This is clearly illustrated in Table 6.

From this table it will be seen that both the anode erosion and the total erosion increase in the following order: cyclohexane, paraffins, aromatics. The large rise in total erosion in this series is due to the cathode. With the aromatics, especially benzene, there is a marked rise in the wear ratio. Thus, the above data indicate that there is a redistribution of the energy of the discharge between the electrodes as we go from saturated hydrocarbons to aromatics.

Our results support the conclusions of Bruma and Magat [4], who inferred from technical tests that the productivity of the electroerosion process can be increased by using aromatic fractions of petroleum products as the interelectrode medium.

The specific total erosion and anode erosion indices indicate that the use of aromatic hydrocarbons in the process will often lead to an improvement in the utilization of energy. The specific yields of eroded material are often higher for aromatics than for paraffins and cyclohexane.

SUMMARY

1. The physicochemical properties of single hydrocarbons used as interelectrode media have a marked effect on the electroerosion of metals.

2. Empirical equations are calculated for the decomposition of the hydrocarbons. From thermodynamic calculations and oscillograms it is established that the energy expended on chemical processes ranges over wide intervals — for paraffins and cyclohexane, 5.52-27.92%; for benzene and o-xylene, 1.73-6.52%; and for tetralin, 10.02-11.36% of the total pulse energy.

3. The anode erosion and the total electrode wear increase in the following order: cyclohexane, paraffins, aromatics.

LITERATURE CITED

1. Pechuro, N. S., A. N. Merkur'ev, É. Ya. Grodzinskii, and N. I. Sokolova, Symposium: Problems in the Electrical Treatment of Materials, No. 2, Izd. Akad. Nauk SSSR, Moscow (1960), p. 14.
2. Steininger, Z., and T. Dolinsky, Mechanic, No. 4:183 (1962).
3. Merkur'ev, A. N., N. S. Pechuro, L. A. Roiter, V. I. God'din, and O. Yu. Pesin, Symposium: Electron Treatment of Materials, Izd. Akad. Nauk MoldSSR, Kishinev (1965), p. 35.
4. Magat, M., and M. Bruma, French Patent, Cl. B23 pH05b, No. 1254441 (1961).

DECOMPOSITION OF ORGANIC LIQUIDS, USED
AS INTERELECTRODE MEDIA IN THE ELECTROEROSION
OF METALS, IN A CONDENSED DISCHARGE

A. N. Merkur'ev, O. Yu. Pesin, and N. S. Pechuro

The industrial introduction of new steels and alloys has stimulated the search for new methods of working them. One such method, which has been extensively developed in recent years, is electroerosion, which is based on the action of certain types of electric discharges on the workpiece. Research in this field has established the theoretical bases of the process, and enabled us to devise a number of techniques and operations and to design special equipment for them.

So far, the least-studied aspects have been the role of the interelectrode medium's composition and the course of its decomposition under the action of various types of electric discharge. These phenomena have much in common with the electrocracking of liquid organic feedstock for producing acetylene-rich gases. However, the decomposition of the interelectrode medium is substantially affected by certain special features of electroerosion, primarily the special power sources, the metal electrodes, and the absence of forced removal of the gaseous decomposition products from the work zone.

Previous research has been aimed at the selection of interelectrode media for specific technological operations. Those mainly chosen have been commercial petroleum products. Only a few authors refer to the use of alcohols and deionized water.

In this article we shall give the results of experiments aimed at establishing how individual organic compounds, used as interelectrode media, decompose when they are acted on by condensed discharges. We paid special attention to the yields and compositions of the decomposition products, and to the proportion of the discharge's energy which is expended on chemical reactions. The hydrocarbons chosen were representatives of the main classes of organic compounds found in various petroleum products. These latter are, of course, the most commonly used interelectrode media.

In the experiments we used a laboratory test rig for electroerosional drilling, previously described in [1]. The workpieces were steel (St-3) blanks (50 x 50 x 20 mm). The electrode instrument was a brass (LS-59) cylinder (d = 20 mm, H = 30 mm). Representative paraffins, naphthenes, and aromatic hydrocarbons were subjected to decomposition.

The power source was an RC oscillator. The mode of operation was governed by the capacitance C, and the principal parameters of the discharges are listed in Table 1. From it we see that the current strength, duration, and energy of both the forward and reverse halfwaves depend markedly on the capacitance; there is little variation in the voltage. In addition, oscillograms showed that the energy of the reverse halfwave is about 10-15% of that of the forward halfwave. In subsequent calculations involving the pulse energy, we therefore took the corresponding values for the forward halfwave. In addition, the energies of the discharges were practically independent of the nature of the medium.

120

TABLE 1. Principal Parameters of Condensed Discharges

C, μF	Forward halfwave				Reverse halfwave			
	U, e	I, a	τ, μsec	W, J	U, e	I, a	τ, μsec	W, J
10	41.2	51.8	32.5	0.033	29.0	16.0	30.0	0.001
70	45.8	189.5	75.6	0.314	30.2	58.3	51.2	0.043
380	55.5	442.0	200.0	2.360	36.4	121.3	114.0	0.280

TABLE 2. Yields of Gaseous Decomposition Products (in 10^2 ml/pulse)

C, μF	n-Heptane	n-Octane	n-Decane	n-Tetradecane	Benzene	o-Xylene	Tetralin	Cyclohexane
10	0.29	0.29	0.24	0.28	0.16	0.16	0.20	0.15
70	2.24	2.30	2.46	2.74	1.14	1.19	1.72	1.63
380	10.70	9.10	8.60	10.90	7.56	8.74	10.95	9.60

TABLE 3. Composition (in vol.%) of Gaseous Products Formed by the
Decomposition of Various Individual Compounds

Medium	C, μF	H_2	CH_4	C_2H_4	C_2H_4	C_3H_4	C_2H_2
n-Heptane	10	70.84	3.03	0.22	7.52	0.29	18.10
	70	62.87	3.74	0.35	10.41	0.48	22.15
	380	60.65	4.58	0.46	12.90	0.91	20.50
n-Octane	10	67.52	4.30	0.30	10.10	0.48	17.30
	70	68.18	3.85	0.22	9.20	0.45	18.10
	380	58.17	4.06	0.43	14.00	0.74	22.60
n-Decane	10	70.54	4.08	0.16	8.81	0.31	16.10
	70	58.99	4.39	0.37	14.00	0.45	21.80
	380	55.92	3.60	0.38	16.60	0.90	22.60
n-Tetradecane	10	00.74	2.31	0.40	16.10	0.85	19.60
	70	59.00	3.83	0.23	13.70	0.64	22.60
	380	55.72	3.60	0.41	16.10	0.77	23.40
Benzene	10	87.60	1.40	—	1.40	—	9.60
	70	87.30	1.00	—	1.10	—	10.60
	380	84.60	—	—	1.90	—	13.50
o-Xylene	10	83.10	3.05	—	0.73	—	13.12
	70	83.70	3.71	—	0.79	—	11.80
	380	82.80	4.56	—	1.14	—	12.10
Tetralin	10	80.35	3.33	—	3.52	—	12.80
	70	73.88	3.28	—	4.66	—	18.18
	380	68.59	3.63	—	5.85	—	21.93
Cyclohexane	10	81.05	0.97	—	4.33	—	13.65
	70	76.57	1.94	—	5.19	—	16.30
	380	69.00	1.70	—	7.40	—	21.90

Table 2 shows that the yield of gaseous decomposition products (in ml/pulse) depends markedly on the nature of the initial medium. It was found that the maximum quantity of gas is formed by decomposition of paraffins, the minimum from aromatic compounds.

We established the following relations between the nature of the compound and the composition of the gas (Table 3). Increasing the chain length of the hydrocarbon from 7 to 14 atoms (in the normal paraffin

A. N. MERKUR'EV, O. YU. PESIN, AND N. S. PECHURO

TABLE 4. Empirical Equations for the Decomposition of Individual Hydrocarbons in a Condensed Discharge

Medium	$c, \mu F$	Equation of decomposition reaction	Thermal effect of reaction, kcal/mole
n-Heptane	10	$C_7H_{16} = 5.074\ H_2 + 0.22\ CH_4 + 0.016\ C_2H_6 + 0.54\ C_2H_4 + 0.021\ C_3H_6 + 1.29\ C_2H_2 + 3.025\ C$	117.40
	70	$C_7H_{16} = 4.34\ H_2 + 0.26\ CH_4 + 0.02\ C_2H_6 + 0.72\ C_2H_4 + 0.03\ C_3H_6 + 1.53\ C_2H_2 + 2.09\ C$	132.04
	380	$C_7H_{16} = 4.04\ H_2 + 0.304\ CH_4 + 0.029\ C_2H_6 + 0.86\ C_2H_4 + 0.06\ C_3H_6 + 1.36\ C_2H_2 + 2.017\ C$	123.52
n-Octane	10	$C_8H_{18} = 5.25\ H_2 + 0.33\ CH_4 + 0.023\ C_2H_6 + 0.785\ C_2H_4 + 0.037\ C_3H_6 + 1.34\ C_2H_2 + 3.263\ C$	142.80
	70	$C_8H_{18} = 5.39\ H_2 + 0.30\ CH_4 + 0.02\ C_2H_6 + 0.72\ C_2H_4 + 0.03\ C_3H_6 + 1.42\ C_2H_2 + 3.59\ C$	158.72
	380	$C_8H_{18} = 4.32\ H_2 + 0.33\ CH_4 + 0.03\ C_2H_6 + 1.04\ C_2H_4 + 0.055\ C_3H_6 + 1.69\ C_2H_2 + 1.985\ C$	141.41
n-Decane	10	$C_{10}H_{22} = 6.81\ H_2 + 0.39\ CH_4 + 0.015\ C_2H_6 + 0.85\ C_2H_4 + 0.03\ C_3H_6 + 1.55\ C_2H_2 + 4.68\ C$	159.60
	70	$C_{10}H_{22} = 5.42\ H_2 + 0.40\ CH_4 + 0.03\ C_2H_6 + 1.28\ C_2H_4 + 0.04\ C_3H_6 + 1.99\ C_2H_2 + 2.87\ C$	177.92
	380	$C_{10}H_{22} = 5.05\ H_2 + 0.32\ CH_4 + 0.03\ C_2H_6 + 1.48\ C_2H_4 + 0.08\ C_3H_6 + 2.02\ C_2H_2 + 2.38\ C$	182.14
n-Tetradecane	10	$C_{14}H_{30} = 7.54\ H_2 + 0.29\ CH_4 + 0.05\ C_2H_6 + 2.00\ C_2H_4 + 0.10\ C_3H_6 + 2.43\ C_2H_2 + 4.45\ C$	229.81
	70	$C_{14}H_{30} = 7.4\ H_2 + 0.48\ CH_4 + 0.03\ C_2H_6 + 1.725\ C_2H_4 + 0.08\ C_3H_6 + 2.84\ C_2H_2 + 4.09\ C$	241.04
	380	$C_{14}H_{30} = 6.85\ H_2 + 0.44\ CH_4 + 0.05\ C_2H_6 + 1.98\ C_2H_4 + 0.09\ C_3H_6 + 2.87\ C_2H_2 + 3.49\ C$	250.92
Benzene	10	$C_6H_6 = 2.56\ H_2 + 0.04\ CH_4 + 0.04\ C_2H_4 + 0.28\ C_2H_2 + 5.32\ C$	3.24
	70	$C_6H_6 = 2.56\ H_2 + 0.03\ CH_4 + 0.03\ C_2H_4 + 0.311\ C_2H_2 + 5.28\ C$	5.02
	380	$C_6H_6 = 2.50\ H_2 + 0.56\ C_2H_4 + 0.391\ C_2H_2 + 5.106\ C$	10.17
o-Xylene	10	$C_8H_{10} = 3.97\ H_2 + 0.147\ CH_4 + 0.035\ C_2H_4 + 0.663\ C_2H_2 + 6.457\ C$	24.32
	70	$C_8H_{10} = 4.00\ H_2 + 0.18\ CH_4 + 0.04\ C_2H_4 + 0.56\ C_2H_2 + 6.62\ C$	21.78
	380	$C_8H_{10} = 3.89\ H_2 + 0.216\ CH_4 + 0.065\ C_2H_4 + 0.57\ C_2H_2 + 6.536\ C$	29.58
Tetralin	10	$C_{10}H_{12} = 4.51\ H_2 + 0.19\ CH_4 + 0.20\ C_2H_4 + 0.72\ C_2H_2 + 7.97\ C$	55.50
	70	$C_{10}H_{12} = 4.11\ H_2 + 0.18\ CH_4 + 0.26\ C_2H_4 + 1.01\ C_2H_2 + 7.28\ C$	72.97
	380	$C_{10}H_{12} = 3.76\ H_2 + 0.20\ CH_4 + 0.32\ C_2H_4 + 1.21\ C_2H_2 + 6.74\ C$	81.80
Cyclohexane	10	$C_6H_{12} = 4.62\ H_2 + 0.055\ CH_4 + 0.248\ C_2H_4 + 1.782\ C_2H_2 + 3.885\ C$	81.72
	70	$C_6H_{12} = 4.28\ H_2 + 0.11\ CH_4 + 0.293\ C_2H_4 + 0.915\ C_2H_2 + 3.474\ C$	101.58
	380	$C_6H_{12} = 3.80\ H_2 + 0.10\ CH_4 + 0.40\ C_2H_4 + 1.20\ C_2H_2 + 2.70\ C$	88.41

TABLE 5. Some Theoretical Characteristics of the Decomposition of Various Interelectrode Media in a Condensed Discharge

	C, µF	n-Heptane	n-Octane	n-Decane	n-Tetra-decane	Benzene	o-Xylene	Tetralin	Cyclo-hexane
(Soot yield)·10^5, g/pulse	10	0.07	0.064	0.063	0.054	0.15	0.123	0.16	0.06
	70	0.34	0.564	0.418	0.484	1.113	0.89	1.11	0.548
	380	1.70	1.32	1.12	1.66	7.09	5.01	7.95	2.65
(Quantity of medium decomposed)·10^5, g/pulse	10	0.189	0.186	0.159	0.196	0.19	0.16	0.22	0.101
	70	1.35	1.49	1.73	1.94	1.369	1.18	1.73	1.11
	380	7.05	6.55	5.64	7.9	9.04	6.80	12.96	6.60
(Amount of energy expended on chemical processes)·10^2, J	10	0.93	0.97	0.746	0.95	0.028	0.166	0.388	0.41
	70	7.44	8.65	9.0	10.06	0.368	1.02	3.99	4.85
	380	36.4	33.7	29.8	41.7	4.92	7.9	33.5	33.3
Proportion of energy of pulse expended on chemical processes, %	10	30.00	28.5	21.31	31.66	0.86	4.54	12.51	14.1
	70	25.40	24.7	28.8	28.34	1.172	3.04	13.90	17.3
	380	15.30	16.20	14.05	17.7	1.92	2.92	15.02	14.7

TABLE 6. Erosion of Electrodes in Various Media

	C, µF	n-Heptane	n-Octane	n-Decane	n-Tetra-decane	Benzene	o-Xylene	Tetralin	Cyclo-hexane
(Total erosion)·10^5, g/pulse	10	0.129	0.152	0.120	0.170	0.156	0.148	0.185	0.132
	70	1.690	1.830	1.500	1.770	2.720	2.380	3.010	1.895
	380	14.930	11.600	10.380	12.400	19.780	18.710	25.100	14.200
(Erosion of workpiece)·10^5, g/pulse	10	0.057	0.071	0.052	0.087	0.078	0.071	0.098	0.066
	70	0.880	0.990	0.710	1.000	1.630	1.400	1.830	1.045
	380	7.830	6.800	5.280	7.100	11.000	9.670	13.800	7.780
(Erosion of electrode)·10^5, g/pulse	10	0.072	0.081	0.068	0.083	0.078	0.077	0.087	0.066
	70	0.810	0.840	0.790	0.770	1.090	0.980	1.180	0.850
	380	7.100	4.800	5.100	5.300	8.780	9.040	11.300	6.420

series) has practically no effect on the gas composition. The gas obtained by decomposing cyclohexane and aromatic compounds has low acetylene and ethylene contents, and contains no ethane or propylene but large amounts of hydrogen. The presence of two methyl groups in the o-xylene molecule is apparently the reason why this compound gives more methane than does benzene. In addition, we may note that more stringent conditions (i.e., higher capacitances) are as a rule accompanied by a somewhat increased content of acetylene and ethylene in the product gas.

Our data on the yield and composition of the gas enabled us to calculate the overall equations for the decomposition of the compounds (Table 4) and to determine the proportion of the pulse energy which was expended on chemical processes (Table 5).

On comparing the equations in Table 4, we can discern several regularities. For example, for the paraffins, the amounts of hydrogen, ethylene, and acetylene formed per molecule of feedstock increase with the molecular weight of the initial product. From the decomposition of the aromatic compounds benzene and o-xylene, we find a lower hydrogen yield than from the paraffins. For o-xylene, tetralin, and cyclohexane, the amount of hydrogen formed per molecule of initial compound is close to the figure for the paraffins, but the amount of acetylene formed remains somewhat less.

An important characteristic of the process is the amount of soot formed by the decomposition of each medium. The formation of soot and its gradual accumulation in the liquid impairs removal of the metallic erosion products, and leads to coking up of the interelectrode gap and reduction in the process productivity. It is rather difficult to determine the soot formation experimentally by compiling mass balances, because the quantity of soot formed in each experiment is so small, and because we do not presently have a reliable method of quantitatively determining the metallic electrode erosion products. Nevertheless, the empirical equations for the decomposition enable us to determine the required value quite easily by calculation [2].

The data in Table 4 show that the maximum amount of carbon is formed by the decomposition of the aromatic compounds. A similar relation was obtained for the soot yield per pulse (see Table 5).

The expenditure of discharge energy on decomposition of the medium is characterized by the thermal effects of the above decomposition reactions. From the data in Table 5 we find that in the normal paraffin series the thermal effect of the reaction increases with the length of the molecule. In the decomposition of cyclohexane and the aromatics, on the other hand, it decreases. The proportion η of the pulse energy expended on chemical processes is determined by the course of the decomposition, the thermal effect of the reaction, and the molecular structure of the initial compound; it is greatest for the paraffins, and least for the aromatics. In addition, as a rule, η falls with increasing pulse energy.

We also measured the electrode erosion in relation to the medium used.

From Table 6 it will be seen that when paraffins are used as the medium, there is little variation in the total erosion. The maximum erosion is obtained with tetralin. Similar laws are obeyed by the erosion of the workpiece and electrode. Table 6 also shows that the medium influences the amount of erosion more markedly in stringent conditions.

Thus, from the data in Tables 5 and 6, we can infer that the aromatic compounds cause greater erosion than the paraffins because the former cause a lower fraction of the discharge energy to be expended on chemical processes.

SUMMARY

1. We have found the yields and compositions of the gaseous products formed by decomposition of single hydrocarbons, which are used as interelectrode media in electroerosion, in a condensed discharge.

2. We have drawn up empirical equations for the decomposition reaction and have calculated the proportion of the discharge's energy which is expended on chemical reactions.

3. We have elucidated how the erosion of the electrodes depends on the nature of the interelectrode medium.

LITERATURE CITED

1. Merkur'ev, A. N., N. S. Pechuro, L. A. Roiter, V. I. Gol'din, and O. Yu. Pesin, Symposium: Electron Processing of Materials, Izd. Akad. Nauk MoldSSR, Kishinev (1965), p. 35.
2. Pechuro, N. S., É. Ya. Grodzinskii, and O. Yu. Pesin, Symposium: Problems in the Electrical Processing of Materials, No. 4, Izd. Akad. Nauk SSSR, Moscow (1962), p. 192.

DECOMPOSITION OF LIQUID PETROLEUM PRODUCTS
UNDER PRESSURE IN A HIGH-VOLTAGE ARC

N. S. Pechuro, A. T. Soldatenkov, A. N. Merkur'ev, and V. M. Chizhov

Up to the present time, research on the decomposition of liquid organic feedstocks in electrical discharges for the purpose of obtaining acetylene-rich gases has been carried out at normal or reduced pressure. When the pressure is reduced to 40-100 mm Hg, there is an increase of 5-7 vol.% in the acetylene concentration of the gas [1, 2]. However, practical realization of electrocracking under reduced pressure involves various technical difficulties. In addition, the extra energy expended on maintaining the vacuum must inevitably lead to increased costs.

In our opinion, electrocracking under increased pressure may be of considerable interest, because the gas formed can, without additional compression, be fed from the reactor to the purification, separation, and further processing stages, thus greatly simplifying the process technology.

In this article we shall give the results of research on the decomposition of two petroleum products in a high-voltage arc at 1-5 atmospheres. We used a laboratory apparatus which is shown diagrammatically in Fig. 1.

The original medium is decomposed in reactor 1. The gaseous products, mixed with liquid circulating medium, arrive at the separating chamber 2. The gas then passes through a filter 3, for removal of solids and liquids, a throttle 4, and a gas meter 5. The liquid products are drawn off from the lower part of the separating chamber 2, and pass through a condenser 6 and a gearwheel-type pump 7, and then return to the reactor for decomposition. The gas pressure is measured by dial gauge 8 and manometer 9. Gas is removed for analysis via pipet 10.

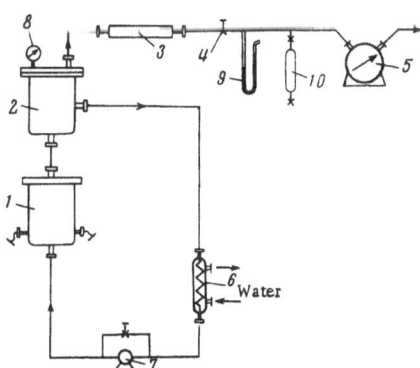

Fig. 1. Diagram of apparatus.

The reactor (Fig. 2) is made of thick-walled seamless steel tubing, and is cylindrical (d = 100 mm, H = 140 mm). A high-voltage arc, in which the feedstock is decomposed, is struck between the upper (movable) and lower (stationary) electrodes.

The feedstock enters via the lower electrode, which is hollow. The mixture of liquid and gaseous products is led out via a side tube in the upper lid. The high-voltage arc is powered from a "VT" transformer (U_{nl} = 30 kV, I_d = 70 mA).

TABLE 1. Characteristics of Initial Petroleum Products

	Petroleum product 1	Petroleum product 2
Molecular weight	153.0	256.5
Ratio of carbon to hydrogen by weight, C/H ...	6.15	6.70
Density, ρ_{20}^{20}, g/cm^3	0.7880	0.8980
Refractive index, n_D^{20}	1.4408	1.4964
Viscosity (Engler)	1.093	3.730
Iodine number (Kaufmann)	9.7	22.8
Total sulfonatables (Kattwinkel).	28.0	33.5
Carbon residue, %	—	0.05
Flash point, °C	49.5	153.0
Ignition point, °C	53.0	173.5
Ibp, °C	150	81 (4 mm Hg)
Fbp, °C	240	200 (4 mm Hg)

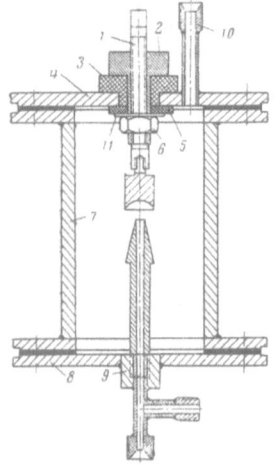

Fig. 2. Reactor. 1) Movable electrode; 2) plug; 3) packing; 4) upper lid; 5) packing; 6) nut; 7) casing; 8) lower lid; 9) conical electrode; 10) connecting tube; 11) washer.

The raw materials were petroleum products Nos. 1 and 2; their chief physicochemical properties are listed in Table 1.

As well as pressure, we studied the effects of various other factors (temperature and linear inlet velocity of the feedstock).

As an example of the decomposition of petroleum product No. 1, we found (Table 2) that with fixed temperatures (30 and 100°C) and constant feedstock velocities (0.9 m/sec), an increase in the pressure leads to an increased gas yield, the acetylene concentration decreasing from 29.2 to 24.1 vol.% (at 30°C), or from 32.5 to 27.0 vol.% (at 100°C).

The methane and ethylene contents increase somewhat with rise of pressure. In these cases, despite the reduction in volume concentration of acetylene in the gas, the absolute acetylene yield rises somewhat, being 2.50 liters/h at 1 atm and 2.65 liters/h at 5 atm (100°C).

Thus, when the temperature of the medium increases from 30 to 100°C, we observe an increase in gas yield and a rise in the concentrations of acetylene and ethylene in the gas. This is apparently due to the fact that, when the pressure rises, there is an increase in selective solution of some of the heavier gas components (acetylene, ethylene, etc.) in the liquid feedstock, and these components are desorbed when the temperature rises, so that their concentrations in the gas increase.

From Table 3 we see that the physicochemical properties of the petroleum products have little influence on the acetylene content of the gas; however, there is a considerable variation in the concentration of lower olefins, which decreases when we work with heavier types of feedstock.

As an example of the decomposition of petroleum product No. 1, we compiled a mass balance for the process and calculated the main parameters (Table 4). These were found from the carbon and hydrogen balances by formulas given in [3]:

TABLE 2. Yield and Composition of Gas Versus Pressure and Temperature
Petroleum Product No. 1; Circulation Velocity of Feedstock 0.9 m/sec

Temp., °C	Pressure, atm	Gas yield liter/h	Gas composition, vol.%						Abs. C_2H_2 yield, liter/h
			H_2	CH_4	C_2H_6	C_2H_4	C_3H_6	C_2H_2	
30	1	7.4	58.0	5.1	0.4	5.9	1.4	29.2	2.16
	2	7.6	56.7	5.1	0.5	8.4	1.7	27.6	2.10
	3	8.1	57.5	5.6	0.5	8.4	1.5	26.5	2.15
	4	8.7	57.9	7.3	0.6	8.2	1.6	24.4	2.12
	5	9.2	56.3	8.6	0.6	9.4	1.0	24.1	2.22
100	1	7.7	50.9	4.9	1.0	8.7	2.0	32.5	2.50
	2	8.0	52.8	5.2	1.1	9.2	1.2	30.5	2.44
	3	8.5	52.0	5.3	1.1	11.0	1.6	29.0	2.47
	4	9.3	53.2	5.9	0.8	11.0	1.8	27.3	2.54
	5	9.8	53.9	5.6	0.8	10.8	1.9	27.0	2.65

TABLE 3. Gas Composition (vol.%) Versus Nature of Initial Feedstock and Pressure in Reactor
Circulation Velocity 2.0 m/sec, 100°C

Pressure, atm	H_2	CH_4	C_2H_6	C_2H_4	C_3H_6	C_2H_2	Pressure, atm	H_2	CH_4	C_2H_6	C_2H_4	C_3H_6	C_2H_2
	Petroleum Product No. 1							Petroleum Product No. 2					
1	52.1	4.1	0.4	8.6	1.6	33.2	1	57.4	3.7	0.4	4.6	0.6	33.3
2	50.6	5.6	1.3	9.3	2.0	31.2	2	56.7	5.5	0.5	5.5	0.7	31.1
3	49.8	5.9	1.1	10.1	2.0	31.1	3	55.7	6.3	0.5	6.6	0.8	30.1
4	51.5	5.9	0.8	11.0	1.5	29.3	4	55.4	6.3	0.6	7.5	0.8	29.4
5	50.8	6.0	0.6	11.5	1.9	29.2	5	53.6	7.6	0.7	8.4	1.0	28.7

TABLE 4. Mass Balance for Decomposition of Petroleum Product No. 1 in a High-Voltage Arc
Circulation Velocity 2.0 m/sec, 100°C

Pressure, atm	Feedstock decomposed, g	Obtained, g		Yields of gas components, g/100 g				Consumption of feedstock, kg/m³ (NTP) C_2H_2	Gas yield, g/100 g	Soot yield, g/100 g
		gas	soot	C_2H_2	C_nH_{2n}	H_2	C_nH_{2n+2}			
1	7.46	6.04	1.42	51.60	18.50	6.23	4.67	2.24	81.0	19.0
3	7.79	6.27	1.52	46.30	21.10	5.78	7.32	2.49	80.5	19.5
5	7.85	6.15	1.70	43.20	22.48	5.80	6.82	2.66	78.3	21.7

$$P^s_{fin} = V \left[K_{init} 0.09 \left(c_{H_2} + c_{C_2H_2} + 2c_{C_2H_4} + 2c_{CH_4} \right) - 0.535 \left(2c_{C_2H_2} + 2c_{C_2H_4} + c_{CH_4} \right) \right];$$
$$N = P^s_{fin} + V\gamma_g,$$

where P^s_{fin} is the absolute yield of soot in g/h; V is the absolute yield of gas in liters/h; K_{init} is the ratio of carbon to hydrogen by weight in the initial products, c_{H_2}, $c_{C_2H_2}$, $c_{C_2H_4}$, and c_{CH_4} are the volume concentrations of the components in the cracking gas; N is the amount of feedstock decomposed in d/h; and γ_g is the weight of 1 liter of cracking gas at NTP.

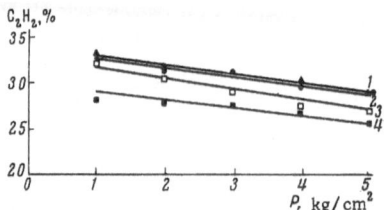

Fig. 3. Concentration of acetylene versus pressure and flow rate
of petroleum product No. 1. Flow rates (m/sec): 1) 7.4; 2) 2.0;
3) 0.9; 4) no circulation.

As will be seen from Table 4, an increase in pressure is accompanied by a reduction in gas yield, an increase in soot yield, and an increase in the feedstock consumed for the production of 1 m^3(NTP) of acetylene.

We also found that the acetylene yield depends on the linear velocity of the feedstock entering the reaction zone from the electrode.

Our results are plotted in Fig. 3. As the linear velocity of the feedstock increases, the acetylene concentration in the gas increases throughout the range of pressures studied (up to 5 atm). This is apparently due to an improvement in the conditions of formation, separation, and quenching of the gaseous products. This is indirectly confirmed by data from high-speed cine films of the decomposition of liquid organic feedstock under 1 or 5 atm (SKS-1 camera, 3000 frames/sec).

Figure 4a shows cine photographs illustrating the formation and development of a gas bubble at normal pressure. The first frame (top) was taken 2 msec after the high-voltage arc ceased. The irregularly shaped gas bubble gradually expands and partly emerges from the reaction zone. However, since there is no circulation of the medium, and since hydraulic shock (such as we find in more powerful electrical discharges) is absent, the gas bubble moves very slowly. Thus, most of the subsequent discharges occur in the gas phase, causing further decomposition of compounds formed in the previous pulse (second frame). The following frames depict the development of the gas bubble 2 and 3 msec after the end of the discharge.

An altogether different picture is obtained when the gas bubble forms and develops under pressure. In Fig. 4b (top frame) we see a gas bubble between the current electrodes; this is due to decomposition of the medium by the preceding pulse (the process is again carried out without circulation of the liquid products). The bubble here is much smaller, although the pulse energy is practically the same. As a rule, in this case breakdown occurs in the gas phase. The discharge is brighter (second frame), and after it has ceased, we see between the electrodes not a gradually spreading gas cloud of irregular shape, but a spherical bubble (or bubbles) which slowly descends the electrodes toward the lower part of the reactor. Here several bubbles unite into a larger one which rises to the upper part of the reactor. The following frames represent the process 1 and 3 msec after the end of the discharge.

We can easily guess that the decrease in volume of the gas bubble and the consequent reduction in its surface of contact with the liquid phase must lead to a deterioration in the quenching of the gaseous products, and this is the cause of the reduced acetylene concentration.

In conclusion, we may state that the decomposition of liquid organic feedstock in a high-voltage arc discharge under pressure gives a gas with a high acetylene content (27-29 vol.%), which can be passed on without additional compression to the purification, separation, and other stages.

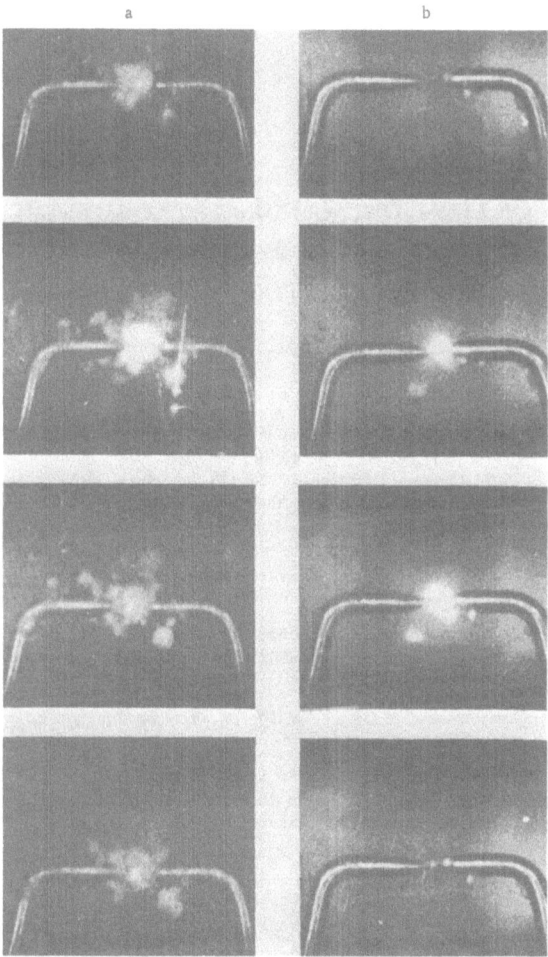

Fig. 4. Frames from cine film of process. a) 1 atm; b) 5 atm.

LITERATURE CITED

1. Kroepelin, H., US Patent No. 2878177 (1957).
2. LaRoi, E., and Hutchings, US Patent No. 2790759 (1957).
3. Pechuro, N. S., É. Ya. Grodzinskii, and O. Yu. Pesin, Symposium: Problems in Electrical Processing of Materials, No. 4, Izd. Akad. Nauk SSSR, Moscow (1962), p. 192.

THE ACTION OF LOW-VOLTAGE NONSTATIONARY ELECTRICAL
DISCHARGES ON LIQUID-HYDROCARBONS
DURING ELECTROCRACKING

E. K. Starostin, I. E. Bulin, and N. S. Pechuro

Most research on the decomposition of liquid organic substances in electrical discharges for the purpose of manufacturing acetylene has been concerned mainly with the design of the plant, the compositions and yields of the products from various types of feedstock, and the effects of the electrical characteristics on the yields and consumptions. Some interest also attaches to the changes which occur in the properties of liquids acted on by various types of electrical discharges. In most of the work so far done in this field [1-4], attention has been paid primarily to such properties as density, molecular weight, refractive index, fractionation properties, iodine number, total sulfonatables, etc. However, there are no data on the changes in the quantitative proportions of individual components in the hydrocarbon mixture.

In this article we describe results obtained by decomposing artificial organic mixtures in low-voltage nonstationary electrical discharges, with the aim of establishing the quantitative behavior of the relative component concentrations.

We used a laboratory reactor shown in Fig. 1. The apparatus comprised the reactor, a condenser system, and the necessary electrical metering devices.

The reactor 1 was cylindrical, made of glass with a cooling jacket 2. Its lower conical section contained the graphite electrodes 10. The lid 4 was used to put in the raw material and the intermediate current-carrying contact, and to take samples of the liquid products.

The intermediate current-carrying contact 3 was a graphite sphere (d = 7-9 mm) which, on making contact with the electrodes 10, completed the circuit. Gas formed by decomposition of the liquid products drove out the sphere so that short circuiting was prevented.

Liquid products partially entrained by the gas current were condensed in condenser 5 and fell back into reactor 1. Further trapping of volatile components of the initial hydrocarbon mixture occurred in trap 6, cooled with dry ice.

The gaseous products, after passing through condenser 5 and trap 6, were collected in gasometer 7. As well as catching the volatile liquid components, trap 6 also partly condensed the gas, which was distilled off after the experiment ended and was collected in gas pipet 8.

The voltages on the electrodes were 60 and 127 V.

In each experiment, about 35-40 g of liquid product was loaded into the reactor; from this, about 16-20 liters of gas were evolved during the experiment. In the decomposition of mixtures of aromatic hydrocarbons, owing to intense soot formation the volume of gas was about 4 liters.

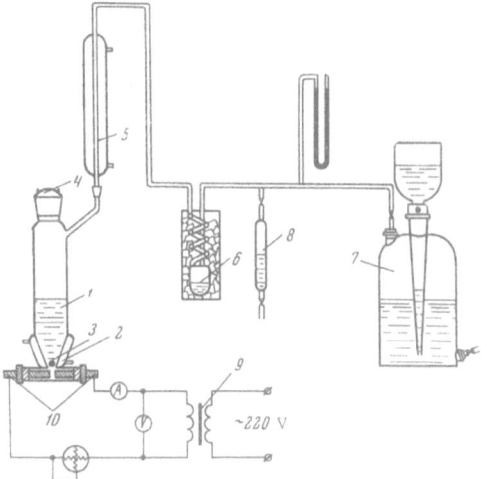

Fig. 1. Diagram of apparatus. 1) Reactor; 2) water jacket; 3) current-carrying contact; 4) reactor lid; 5) vertical condenser; 6) trap; 7) gasometer; 8) gas pipet; 9) transformer; 10) graphite electrodes.

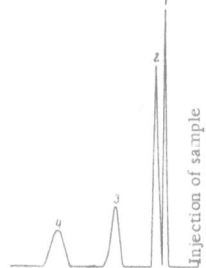

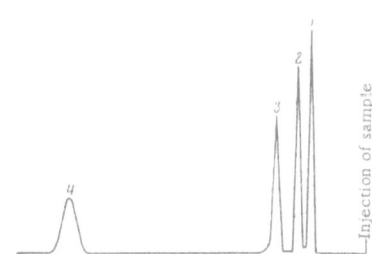

Fig. 2. Chromatogram of separation of mixture of cyclo-olefins at 95°C (stationary phase, squalane). 1) Cyclohexene; 2) methyl cyclohexene; 3) ethyl cyclohexene; 4) propyl cyclohexene.

Fig. 3. Chromatogram of separation of mixture of C_6 hydrocarbons at 70°C (stationary phase, triethyleneglycol butyrate). 1) Hexane; 2) hexene; 3) cyclohexane; 4) benzene.

The soot was filtered off from the liquid, dried, and weighed. The gaseous products were analyzed by gas-adsorption chromatography in an aluminum-oxide column.

The liquid products were analyzed periodically (after the formation of each 4 liters of gas) by gas—liquid chromatography. For analysis of the paraffin and olefin hydrocarbons, the stationary phase was triethyleneglycol butyrate; for aromatic and naphthenic hydrocarbons it was squalane.

TABLE 1. Mass Balances for the Decomposition of Mixtures of Hydrocarbons
in Low-Voltage Nonstationary Discharges

Initial mixture of hydrocarbons	Voltage, V	Wt. of initial mixture, g	Undecomp. liquid products, g	Decomp. prod.			
				gas		soot	
				g	wt.%	g	wt.%
Paraffins.[1*]	60	40.20	26.90	11.78	88.60	1.52	11.40
	127	37.25	26.45	9.10	84.30	1.70	15.70
Olefins[2*]	60	35.04	22.21	10.52	82.00	2.31	18.00
	127	35.35	26.82	6.54	76.70	1.99	23.30
Naphthenes[3*]	60	35.00	23.07	9.60	80.00	2.33	20.00
	127	33.60	25.27	5.92	71.10	2.41	28.90
Cyclo-olefins[4*]	60	32.90	22.82	6.86	68.10	3.22	31.90
	127	33.50	25.47	5.17	64.30	2.86	35.70
Aromatics[5*]	60	35.25	31.30	1.78	45.00	2.17	55.00
	127	35.35	31.24	1.63	40.00	2.48	60.00
Hydrocarbons: C_6[6*]	60	35.20	22.49	9.68	76.20	3.03	23.80
	127	30.50	20.81	6.78	70.00	2.91	30.00
» C_7[7*]	60	35.30	23.43	9.11	76.80	2.76	23.20
	127	35.30	26.62	5.37	61.90	3.31	38.10
» C_8[8*]	60	35.50	22.99	9.73	77.80	2.78	22.10
	127	35.80	27.73	5.37	66.50	2.70	33.50

[1] 18.2% n-hexene, 18.9% n-heptane, 19.6% n-octane, 19.6% n-nonane, 23.7% n-decane.

[2] 20.2% hexene, 19.5% heptene, 19.5% octene, 20% nonene, 20.8% decene.

[3] 25.3% cyclohexane, 24.5% methyl cyclohexane, 24.8% ethyl cyclohexane, 25.4% propyl cyclohexane.

[4] 27% cyclohexene, 26.1% methyl cyclohexene, 25.9% ethyl cyclohexene, 21% propyl cyclohexene.

[5] 18.8% benzene, 20% toluene, 21.3% ethyl benzene, 19.7% propyl benzene, 20.2% butyl benzene.

[6] 25.4% benzene, 28.5% cyclohexane, 21.4% hexene, 24.7% hexane.

[7] 25.8% toluene, 24.7% methyl cyclohexane, 24.7% heptene, 24.8% heptane.

[8] 24.8% ethyl benzene, 25.2% ethyl cyclohexane, 24.3% octene, 25.7% octane.

We studied how carbon-chain length and molecular structure affected the decomposition of artificial mixtures of hydrocarbons of various classes. These mixtures were as follows:

Paraffins: n-hexane, n-heptane, n-octane, n-nonane, and n-decane.

Olefins: hexene, heptene, octene, nonene, and decene.

Aromatics: benzene, toluene, ethyl benzene, propyl benzene, and butyl benzene.

Naphthenes: cyclohexane, methyl cyclohexane, ethyl cyclohexane, and propyl cyclohexane.

Cyclo-olefins: cyclohexene, methyl cyclohexene, ethyl cyclohexene, and propyl cyclohexene.

It was also of interest to study the influence of molecular structure on the electrocracking process. For this purpose we composed mixtures of hydrocarbons with equal numbers of carbon atoms in their molecules:

TABLE 2. Percentage Composition by Volume of Gases Formed by Decomposition
of Mixtures of Hydrocarbons in Low-Voltage Nonstationary Electrical Discharges

Voltage, V	Initial mixture of hydrocarbons	H_2	CH_4	C_2H_4	C_2H_4	C_3H_6	C_2H_2
60	Paraffins	56.6	6.0	0.5	9.8	--	27.1
	Olefins	53.8	4.9	--	9.9	1.2	30.2
	Naphthenes	58.0	5.2	0.2	8.2	0.4	28.0
	Cyclo-olefins	57.0	4.0	--	8.1	1.1	30.8
	Aromatics	62.5	7.7	--	3.1	--	26.7
127	Paraffins	57.0	5.7	0.4	10.0	0.5	26.4
	Olefins	55.6	6.2	0.5	9.2	1.4	27.1
	Naphthenes	59.0	5.9	--	8.2	0.5	26.4
	Cyclo-olefins	55.0	7.0	--	9.0	1.0	28.0
	Aromatics	61.4	7.6	--	3.2	--	27.8
60	Hydrocarbons C_6	52.0	5.1	--	10.6	1.3	31.0
	Hydrocarbons C_7	54.6	4.4	0.3	13.2	1.2	26.3
	Hydrocarbons C_8	56.9	5.0	--	9.3	0.9	27.9
127	Hydrocarbons C_6	57.0	6.7	0.4	7.9	0.7	27.3
	Hydrocarbons C_7	57.2	6.2	--	9.0	0.8	24.8
	Hydrocarbons C_8	57.3	6.7	--	9.7	0.7	25.6

Note: The compositions of the initial mixtures of hydrocarbons are given in the
notes to Table 1.

TABLE 3. Percentage Composition by Weight of Mixtures of Liquid Hydrocarbons with Different Chain Lengths,
Acted on by Nonstationary Electrical Discharges

Components of mixture	Initial mixture, wt.%	60 V						127 V				
		Amount of gas obtained (liters)					Final comp. of products (allowing for condensate in trap)	Amount of gas obtained (liters)				Final comp. of products (allowing for condensate in trap)
		4	8	12	16	20		4	8	12	16	
Paraffins												
n-Hexane . . .	18.2	17.0	18.4	15.3	15.0	12.6	18.0	17.1	16.5	12.7	11.6	17.7
n-Heptane.	18.0	21.1	21.5	19.2	20.0	18.0	19.3	19.2	19.7	19.0	18.6	19.6
n-Octane	19.6	22.7	22.2	22.4	23.1	21.9	21.0	20.2	21.0	22.2	21.8	17.9
n-Nonane	19.6	19.8	18.2	21.5	20.8	21.9	20.0	19.5	19.7	21.9	22.9	22.0
n-Decane	23.7	19.4	19.7	21.6	21.1	24.8	21.5	24.0	23.1	24.2	25.1	22.8
Olefins												
Hexene.	20.2	17.6	15.7	15.8	14.0	—	22.0	17.1	14.3	12.2	—	19.8
Heptene	19.5	21.2	20.3	20.6	19.1	--	20.7	20.4	20.2	20.6	—	21.2
Octene.	19.5	21.7	21.8	22.0	22.2	—	19.6	21.4	22.0	22.8	--	20.0
Nonene.	20.0	20.0	22.7	21.6	24.6	—	20.4	20.7	23.3	23.3	—	20.5
Decene	20.8	19.5	19.5	20.0	20.1	—	17.3	20.4	20.2	21.1	--	18.5
Naphthenes												
Cyclohexane	25.3	25.0	23.4	22.7	22.8	—	27.5	25.3	23.8	21.9	—	26.2
Methyl cyclohexane	24.5	26.0	24.3	22.6	23.6	—	23.7	25.9	26.2	26.1	—	26.2
Ethyl cyclohexane .	24.8	24.5	24.9	25.2	24.7	—	22.5	25.0	25.0	25.5	—	23.4
Propyl cyclohexane	25.4	24.5	27.4	28.5	28.9	—	26.3	23.8	25.0	26.5	—	24.2
Cyclo-olefins												
Cyclohexene	27.0	24.5	25.4	25.6	—	—	28.5	24.9	22.2	—	—	26.2
Methyl cyclohexene.	26.1	27.0	28.2	26.0	--	—	26.0	27.0	27.0	—	—	26.6
Ethyl cyclohexene. . .	25.9	23.5	24.4	24.5	—	—	23.2	24.9	25.0	—	--	24.4
Propyl cyclohexene.	21.0	25.0	22.0	23.9	--	—	22.3	23.2	25.8	—	—	22.8
Aromatics												
Benzene	18.8	19.5	—	—	—	—	19.5	18.5	—	—	—	18.5
Toluene	20.0	20.2	—	--	—	—	20.2	21.0	--	—	—	21.0
Ethyl benzene. . .	21.3	21.8	—	—	—	—	21.8	21.0	—	—	—	21.0
Propyl benzene . .	19.7	20.2	—	—	—	—	20.2	21.0	—	—	—	21.0
Butyl benzene . .	20.2	18.3	—	—	—	--	18.3	18.5	—	—	—	18.5

TABLE 4. Percentage Composition by Weight of Mixtures of Liquid Hydrocarbons of Different Structures, Subjected to Nonstationary Electric Discharges

Components of mixture	Initial mixture, wt.%	60 V					127 V			
		Amount of gas obtained (liters)				Final comp. of products (allowing for condensate in trap)	Amount of gas obtained (liters)			Final comp. of products (allowing for condensate in trap)
		4	8	12	16		4	8	12	
C$_6$ Hydrocarbons										
n-Hexane	20.9	19.6	19.3	19.1	18.4	21.0	23.5	21.9	22.2	23.7
Hexene	24.9	24.7	24.4	24.3	24.8	27.1	21.9	20.0	20.2	22.3
Cyclohexane	28.9	30.3	30.1	30.2	31.2	28.5	28.7	30.1	30.0	28.2
Benzene	25.3	25.1	26.2	26.4	25.6	23.4	25.9	28.0	27.6	25.8
C$_7$ Hydrocarbons										
n-Heptane.	24.8	27.5	28.3	28.1	26.9	27.2	27.4	27.7	26.0	26.1
Heptene	24.7	20.9	20.2	20.6	19.0	19.9	21.8	21.8	21.4	21.5
Methyl cyclohexane . . .	24.7	26.5	27.0	27.3	25.9	26.1	23.9	26.6	27.2	27.2
Toluene	25.8	25.1	24.5	24.0	28.2	26.8	26.9	23.9	25.4	25.2
C$_8$ Hydrocarbons										
n-Octane	25.7	24.4	23.3	22.7	22.4	22.8	24.6	25.5	23.9	24.1
Octene.	24.3	25.4	23.7	22.7	22.5	22.9	23.7	23.9	23.1	23.2
Ethyl cyclohexane.	25.2	27.4	27.6	26.6	27.1	27.3	25.3	26.3	26.9	26.9
Ethyl benzene	24.8	23.8	25.4	28.0	28.0	27.0	26.4	24.3	26.1	25.8

C$_6$: n-hexane, hexene, cyclohexane, benzene.

C$_7$: n-heptane, methyl cyclohexane, toluene, heptene.

C$_8$: n-octane, octene, ethyl cyclohexane, ethyl benzene.

Mass balances for the electrocracking of these mixtures under electrode voltages of 60 and 127 V are listed in Table 1. From these data we see that the yields of decomposition products (gas and soot) depend markedly on the nature of the initial feedstock. We must again emphasize that the maximum amount of soot (55-60% by weight) is formed from the aromatic hydrocarbons, and the minimum (11.0-15.0% by weight) from the paraffins. The olefins, naphthenes, and cyclo-olefins have intermediate soot yields. In the decomposition of the mixtures of hydrocarbons of different classes (paraffins, olefins, naphthenes, and aromatics), the gas yield varies less and is 65-70 wt.% of the feedstock decomposed.

The gas formed by electrocracking the mixtures of hydrocarbons with electrode voltages of 60 or 127 V contains 27-30 vol.% of acetylene, 9-10 vol.% of olefins, 54-60 vol.% of hydrogen, and also some saturated compounds (4-5 vol.%) (Table 2).

The changes in weight ratios occurring in the mixtures of liquid products under the action of nonstationary electrical discharges were assessed by comparing the contents of corresponding components in the initial and final mixtures. For this purpose, after the evolution of each 4 liters of gas the liquid products were analyzed.

Figures 2 and 3 give typical chromatograms of the separation of mixtures of cyclo-olefins and C$_6$ hydrocarbons.

As will be seen from Tables 3 and 4, owing to the different vapor pressures of the components of the mixtures of liquid products, we observed entrainment of the volatile hydrocarbons by the gas current. The final mixtures of liquid feedstock therefore displayed some decrease (4-5%) in the content of volatile hydrocarbons by comparison with the initial feedstock.

However, when account is taken of the condensate in the trap, the final recalculated product composition remains more or less constant regardless of the degree of decomposition.

Thus, in the conditions of our experiments (U_{nl} = 60 or 127 V), hydrocarbons with different chain lengths (from C_6 to C_{10}) in different classes decompose to approximately the same degree when acted on by low-voltage nonstationary electrical discharges.

SUMMARY

When mixtures of hydrocarbons of various classes (paraffins, olefins, naphthenes, and aromatics) with different numbers of carbon atoms in the molecule (C_6 to C_{10}) are decomposed in nonstationary electrical discharges (U_{nl} = 60 or 127 V), all the components of the mixture decompose to approximately the same degree, regardless of the structure and length of the carbon chain.

LITERATURE CITED

1. Linder, E. G., and A. Davis, J. Phys. Chem., 35 : 3649 (1931).
2. Andreev, D. N., Organic Synthesis in Electrical Discharges, Izd. Akad. Nauk SSSR, Moscow-Leningrad (1953).
3. Kroepelin, H., West German Patent No. 1030950 (1958).
4. Pechuro, N. S., A. N. Merkur'ev, and G. A. Grishin, Symposium: Synthesis and Properties of Monomers, Izd. Nauka, Moscow (1964), p. 22.